Review and Evaluation of the VA Enrollee Health Care Projection Model

Katherine M. Harris, James P. Galasso, Christine Eibner

Prepared for the Department of Veterans Affairs

Approved for public release; distribution unlimited

Center for Military Health Policy Research

A JOINT ENDEAVOR OF RAND HEALTH AND THE
RAND NATIONAL DEFENSE RESEARCH INSTITUTE

The research described in this report was sponsored by the Department of Veterans Affairs. The research was conducted in the RAND National Defense Research Institute, a federally funded research and development center sponsored by the Office of the Secretary of Defense, the Joint Staff, the Unified Combatant Commands, the Department of the Navy, the Marine Corps, the defense agencies, and the defense Intelligence Community under Contract W74V8H-06-C-0002.

Library of Congress Cataloging-in-Publication Data

Harris, Katherine M.
Review and evaluation of the VA Enrollee Health Care Projection model / Katherine M. Harris, James P. Galasso, Christine Eibner.
p. cm.
Includes bibliographical references.
ISBN 978-0-8330-4570-6 (pbk. : alk. paper)
1. United States. Dept. of Veterans Affairs—Appropriations and expenditures. 2. Veterans—Medical care—United States—Costs—Forecasting. 3. Veterans—Medical care—United States. I. Galasso, James P. II. Eibner, Christine. III. Title. IV. Title: Review and evaluation of the EHCPM model.

UB357.56.H37 2008
362.11—dc22

2008042243

Published 2008 by the RAND Corporation
1776 Main Street, P.O. Box 2138, Santa Monica, CA 90407-2138
1200 South Hayes Street, Arlington, VA 22202-5050
4570 Fifth Avenue, Suite 600, Pittsburgh, PA 15213-2665
RAND URL: http://www.rand.org/
To order RAND documents or to obtain additional information, contact
Distribution Services: Telephone: (310) 451-7002;
Fax: (310) 451-6915; Email: order@rand.org

Preface

The Veterans' Health Care Eligibility Reform Act of 1996 (PL 104-262) transformed the U.S. Department of Veterans Affairs (VA) from an episodic, inpatient provider to a comprehensive health care provider. It also authorized VA health care for all veterans and mandated that the VA establish a priority-based enrollment system to manage access to care. The legislation requires the Secretary of Veterans Affairs to make an annual determination wether budgeted resources are available to provide timely, high-quality care to all enrolled veterans. Unanticipated changes in the demand for and delivery of health care can yield budget shortfalls or surpluses.

The VA relies on the Enrollee Health Care Projection Model (EHCPM) to project veteran enrollment, enrolled veterans' use of health care services, and the cost of providing these services. Senior VA leadership and regional leadership of the VA's Veterans Integrated Service Networks (VISNs) use projections produced by the model to understand the dynamics of demand for health care services in the VISN service delivery areas and to plan the VA's appropriation request. There is also interest in using the model as a platform for a variety of strategic-planning and policy analysis activities.

The VA asked the RAND Corporation, in conjunction with an independent, senior-level actuary, to review and evaluate the structure of the EHCPM to assess the accuracy, validity, and cost-effectiveness of the model; to identify potential model enhancements; and to assess the risks to the Veterans Health Administration (VHA) posed by the model's reliance on proprietary elements. This monograph will provide

VHA leadership and oversight bodies with insights useful for understanding the EHCPM and its role in the VA's budget and strategic-planning processes.

This study was sponsored by the Department of Veterans Affairs, Veterans Health Administration's Office of the Assistant Deputy Under Secretary for Health for Policy and Planning. It was conducted jointly by RAND Health's Center for Military Health Policy Research and the Forces and Resources Policy Center of the National Defense Research Institute (NDRI). NDRI is a federally funded research and development center sponsored by the Office of the Secretary of Defense, the Joint Staff, the Unified Combatant Commands, the Department of the Navy, the Marine Corps, the defense agencies, and the defense Intelligence Community. More information about RAND is available at www.rand.org.

The principal investigator for the project was Katherine Harris. James P. Galasso, FSA, MAAA, President of Actuarial Services & Financial Modeling, Inc., served as the project actuary. Mr. Galasso fully collaborated with RAND in all aspects of the evaluation process, including reviewing materials, attending briefings, and assisting with writing the final report. The findings expressed in this document represent conclusions reached mutually by RAND and Mr. Galasso.

Contents

Preface iii
Figures ix
Tables xi
Summary xiii
Acknowledgments xxi
Abbreviations xxiii

CHAPTER ONE
Introduction and Background 1
Overview of the Current VA Health Care System 1
Eligibility Reform 3
Budget and Strategic Planning Under Eligibility Reform 4
Opportunities and Challenges of Policy Models 5
Actuarial Models 6
Limitations of VA Data for Forecasting Demand 6
The EHCPM: Description, History, and Uses 8

CHAPTER TWO
Evaluation Methodology 11
Key Activities 11
Criteria for Evaluating Policy Models 13
Validity 13
Accuracy 15
Tractability 17
Transparency 18
Limitations 20

CHAPTER THREE
Model Overview 21
Milliman Health Cost Guidelines 21
Overview of Model Structure 23
Enrollment Projection Model 24
Utilization Projection Model 26
Unit Cost Projection Model 34
Utilization and Cost Trends 41
Budget Reconciliation 43

CHAPTER FOUR
Findings on Model Structure and Validity 45
Enrollment Projection Model 46
Utilization Projection Model 48
Derivation of VA-Specific Utilization Benchmarks 48
Morbidity and Reliance Adjustments for Non–Medicare Eligible Enrollees 49
Benchmarking to Community Management Practices 50
Unit Cost Projection Model 52
Prototype Staffing Model 54
Scenario 1 55
Scenario 2 56
Scenarios 3 and 4 57
Utilization and Cost Trends 57

CHAPTER FIVE
Findings on Model Accuracy 61
Accuracy of the Utilization Projection Model 63
Quality Assurance Procedures 64

CHAPTER SIX
Findings on Tractability and Transparency 65
High-Level Structure 65
Complexity 66
Proprietary Model Elements 66
Documentation 67

CHAPTER SEVEN
Benefits and Risks of the EHCPM 69
Benefits of the Current EHCPM 69
Risks of the Current EHCPM 71

CHAPTER EIGHT
Conclusions 75
EHCPM Use in Short-Term Budget Projection 75
EHCPM Use in Strategic Planning and Policy Analysis 75
Future Modifications to the EHCPM 76

APPENDIXES
A. Results of Validation Studies 79
B. Priority-Level Definitions 87
C. Model Uses Described in "VA Enrollee Health Care Projection Model Training Companion Manual," June 2006 89

References 91

Figures

3.1. High-Level Structure of the EHCPM 24
4.1. Projected Office Visit Expenditures for Priority 2 Males Aged 50–59 Who Enrolled Pre–Eligibility Reform, by Model Type, for Four Fixed-Cost Share and Available-Capacity Scenarios 56

Tables

3.1. Utilization Projection Model Example 28
3.2. Unit-Cost Basis, by Health Care Service Category 36
3.3. Derivation of Detailed Unit Costs Using Office-Visit Example 39
4.1. Cost Structure and Capacity Assumptions Considered Using Prototype Staffing Model 55
A.1. Actual-to-Expected Adjustment Factors, Model Error Ratios, and Projection Error Ratios by Selected Service Types and Selected Enrollee Types 81
A.2. Base Year Modeled to Actual Utilization Relativities by VISN 84

Summary

The Veterans' Health Care Eligibility Reform Act of 1996 significantly expanded the mission of the VA. The reform act vastly increased the types of services offered to VA patients and extended medical coverage to all veterans through a priority-based enrollment system. The VA now operates the largest integrated health care system in the United States. In 2007, the VA had 7.8 million enrollees, served 5.5 million patients, and had a total operating budget of $37.3 billion.

To meet veterans' health care needs under its expanded mission, the VA needs accurate forecasts of future resource requirements so that it can plan and budget accordingly. Developing these estimates is a formidable task because demand for VA resources is variable and difficult to predict. Currently, the VA uses a complex model—the EHCPM—as a planning tool to estimate future demand for medical care and related services among U.S. veterans. The model is maintained and operated by a contractor, Milliman, Inc. The EHCPM is currently used to develop the VA's annual budget request. The VA would like to extend the model's uses to support policy analysis and strategic planning.

The utility of the EHCPM depends on its ability to provide accurate and timely projections of future demands on VA resources. To gauge the model's utility as a budgeting and planning tool, the VA asked RAND and an independent, senior-level actuary to conduct an assessment of the EHCPM. Specifically, RAND was asked to perform two tasks:

- Review the model's key features to determine how effectively they support the VA's budget and planning needs.

- Assess the benefits and risks associated with current model specifications and current contractual arrangements.

Study Methods

To conduct the assessment, RAND, in conjunction with subcontractor Actuarial Services & Financial Modeling, Inc., performed several tasks. We reviewed model documentation; generated and reviewed responses to ad hoc questions about model features and contractual arrangements posed to VA and Milliman; reviewed Milliman's corporate capabilities and background, the qualifications of key Milliman staff, and a summary of hours billed from March 2005 to May 2006; attended a 2-day model training course; visited Milliman headquarters to review proprietary model inputs; participated in a half-day discussion with senior VA staff from the VA Office of the Assistant Deputy Under Secretary for Policy and Planning and the Office of the Chief Financial Officer; reviewed accuracy and validity studies prepared by Milliman; and developed a small, prototype model that replicates EHCPM output under alternative assumptions regarding VA's cost structure.

How the Model Works

The EHCPM model projects total expenditures in any given year by combining output from three subcomponents—enrollment level, utilization rate, and unit cost. These elements are multiplied together for each of 58 medical services for roughly 40,000 enrollee types, or "cells." These cells are defined by age category, by whether enrollment occurred before or after eligibility reforms, by priority level, and by geographic sector. The model applies four types of trend factors to account for general changes in medical costs and the anticipated changes in the efficiency of VA providers. The trend factors are utilization, inflation, intensity of service provision, and a measure of management efficiency (referred to by Milliman as the "degree of community management"). In addition, the model accounts for anticipated changes in veteran morbidity and reliance on the VA health care delivery system, enrollment levels, and enrollment mix. The model's three subcomponent models are described in more detail below.

Enrollment Projection Model

The enrollment model is the simplest of the three subcomponents. It develops projections by applying historic enrollment rates to the forecast veteran population derived from U.S. Census data. Modeled enrollment rates are obtained by age, priority level, geographic sector, and participation in Operation Enduring Freedom or Operation Iraqi Freedom—referred to as *special conflict* status. The projected enrollee population is equal to current enrollment plus new enrollment minus deaths. Although enrollment rates reflect demographic trends in the veteran population, such as shifts in priority level and geographic migration, they do not account for trends in the generosity, availability, and affordability of private-sector health insurance that could lead veterans to enter or leave VA health care.

Utilization Projection Model

Modeled utilization is based on the Milliman Health Cost Guidelines (HCGs), a proprietary set of utilization-rate benchmarks derived from commercial data. The HCGs contain data on utilization for 37 of the 55 EHCPM health service categories. These data are based on a standard fee-for-service benefit package. Milliman applies a complex set of adjustments to the HCG data to reflect the health status of VA enrollees, their reliance on VA, and the relative efficiency of VA facilities. In each projection year, utilization rates are adjusted to account for national trends in health care utilization and VA-specific trends in management efficiency. Because model management trends are calibrated against the local community in which each VA facility operates, projected changes over time in the efficiency of VA practice are implicitly tied to community practices. In a final step, adjusted HCG benchmarks are calibrated to actual VA workload in the model base year to account for differences between the VA and the private sector that are not captured by adjustment. Milliman refers to this as the "Actual to Expected Adjustment." Services without commercial counterparts (such as VA-specific outpatient mental health services, blind rehabilitation, and over-the-counter drugs and supplies) are projected directly from VA workload data.

Unit Cost Projection Model

Average unit cost (that is, the cost of a particular medical service) is derived by allocating VA's base year budget obligation to base year VA workload in each service category. Many of the service categories in the utilization model are developed at a finer level of detail than the service categories defined in the VA's cost accounting system. In these cases, the model relies on calculated relationships between VA cost levels and Medicare-allowable or billed charges to estimate VA unit costs by service category. Inflation and intensity trends are then multiplied by base year average unit costs to project unit costs in any given year.

Results of the Assessment

Our assessment addressed four main questions: (1) Validity—does the model measure what it is intended to measure? (2) Accuracy—how accurately does the model forecast the outcomes it is intended to project? (3) Tractability—how easily can modelers and users understand, and potentially replicate, the model's features, especially its validity and accuracy? (4) Transparency—how clear are the model's assumptions and operating processes? The results of our assessment are summarized below.

How Valid Is the Model?

We conclude that the EHCPM is likely to be valid for short-term budget planning but may not be valid for longer-range planning and policy analysis. The EHCPM represents a substantial improvement over the budgeting methodologies used by the VA in the past for two reasons: (1) The model builds total expenditures from detailed service categories and enrollee types, and (2) it disaggregates enrollment, utilization, and cost components. Thus, the VA can use the current specification to identify factors that drive specific types of expenditures or expenditures for specific types of enrollees. The VA can also use the model to develop more-informed strategies for managing expenditures and allocating budget appropriations. Because it incorporates a wide range of utilization and unit cost parameters for a variety of services and enrollee types, the current model structure can be used to monitor budget execution and performance relative to preestablished benchmarks. Limited accuracy and timeliness in the VA data systems—not limitations in the model structure itself—are the only constraints on the model's utility with respect to these functions.

However, the model could yield misleading results when used for strategic planning and policy analysis. Using the model to inform scenarios beyond the current policy and budgetary environment requires assumptions about a number of factors, including the VA's cost structure; how rapidly the VA can expand its capacity to meet demand; the factors driving enrollment levels; the comparability of patterns of health care use in the VA and commercial sectors; and the relationships among enrollee health status, VA treatment capacity, and enrollees' preferences for treatment in VA facilities. In many cases, a lack of appropriate data or uncertainty about the future makes the assumptions difficult or even impossible to assess.

Modifications to model subcomponents and enhancements of supporting data inputs would likely be required before the model could support a broader range of applications beyond short-term budget planning. These modifications are needed for two reasons. First, under the current specification, short-term utilization projections are tied to VA experience and thus do not measure potential demand independent of the current VA delivery system. Second, the current specification treats the VA's cost structure like that of a fee-for-service payer, such as Medicare or a commercial insurer. Thus, if a substantial proportion of the VA's costs are fixed, projected expenditures will be unrealistic. Fortunately, the model has a flexible, component-based structure under which modifications can be implemented without sacrificing the continuity of budget-planning applications.

How Accurate Is the Model?

The accuracy of the EHCPM is difficult to assess and is thus uncertain. As we discuss below, factors that hinder assessment of model accuracy are, in large part, those that limit model validity. In our view, the most challenging barrier to accuracy stems from the lack of unit cost measures that are independent of the VA's budget allocation. This is because the discretionary nature of the VA's budget complicates the relationship between model projections and actual expenditures. Under a discretionary budget, the VA does not have the authority to spend more than Congress appropriates. If demand for VA services cannot be satisfied under its appropriation, then "actual" expenditures will reflect the constraints inherent in the appropriation and not true demand for VA care, which model developers strive to project. In such a circumstance, accessing the overall accuracy of expenditure forecasting models by comparing projections against an approved budget can be misleading.

Several other factors further complicate an assessment of EHCPM accuracy, over and above those that must be addressed in assessing the accuracy of policy models more generally. Because the VA data systems upon which the model relies and the current structure of the VA benefit are relatively new, opportunities for assessing accuracy against historical utilization experience are limited. Finally, because unmet demand is not observed, it is not possible to compare projected demand with actual total demand. To our knowledge, no specialized analytic methods have been established or ancillary data collected and analyzed for the purpose of assessing accuracy in the context of these challenges.

How Tractable and Transparent Is the Model?

The overall structure of the model is relatively easy to understand. However, this is not true of the model's subcomponents. Tractability and transparency are reduced by the complexity of adjustment algorithms used to set parameters of model subcomponents; the uneven and often incomplete model documentation; reliance on proprietary utilization data and clinical efficiency benchmarks, the quality and appropriateness of which cannot be reviewed by interested parties; and the lack of a process for obtaining independent reviews of the model by outside experts.

Benefits and Risks of Using the EHCPM

Based on our review of model features and an informal assessment of the contractual arrangements under which the EHCPM is developed and maintained, we assessed the benefits and risks to the VA of the current model specification and existing contractual arrangements. Compared to traditional methods, the current specification offers the benefit of a substantially more flexible and detailed platform from which to plan the VA's appropriation request, monitor budget execution, and assess system performance. The main risk to the VA stems from the potential for misleading projections when the model is used to inform future policy and budget decisions. Overall, we find the risks of outsourcing to be low and manageable. The most important risk of outsourcing to consider is the lost opportunity to build institutional knowledge of internal VA staff through day-to-day participation in model-related activities.

Conclusions

While the current EHCPM is useful for short-term budget planning, our review suggests it is of limited utility for planning and policy analysis. To enhance the utility of the model for these activities, the VA might consider modifications to model subcomponents to allow for more-robust forecasting of the demand for and cost of VA care in a changing policy environment. Such modifications are likely to require substantial investments to expand VA's on-going survey efforts and to develop tools for measuring treatment capacity and costs. If such investments are not practical or feasible, the VA may want to investigate simplifications of the current model that draw more exclusively on the VA's own data resources. A simpler model would be more transparent to model constituents and may perform equally well. Under either an enhanced or simplified model, the VA might also consider other improvements, including more-approachable and complete documentation, involvement of a wider range of experts in model development, and periodic review by independent experts.

Acknowledgments

The authors wish to thank Duane Flemming, Barbara Manning, Kathi Patterson, and Merideth Randles for providing materials to and organizing meetings for our study team. We are particularly grateful for their timely response to our questions about the technical aspects of the model as well as how the model is utilized. We are grateful to Susan Hosek, Terri Tanielian, and Jeffrey Wasserman for their guidance in developing and reviewing our findings and recommendations, as well as to Jeanne Ringel and Guy King, who provided valuable comments on an earlier draft of this monograph. Together, their suggestions have greatly improved the final product.

Abbreviations

A/E	actual to expected
ActMod	Actuarial Services & Financial Modeling, Inc.
ARC	Allocation Resource Center
CARES	Capital Asset Realignment for Enhanced Services
CBOC	Community-Based Outpatient Clinic
CMS	Center for Medicare and Medicaid Services
CPDS	Chronic Illness and Disability Payment System
DoCM	degree of community management
DSS	Decision Support System
EHCPM	Enrollee Health Care Projection Model
EPM	Enrollment Projection Model
GAO	General Accounting/Government Accountability Office
FY	fiscal year
HCG	Health Cost Guideline
HSC	health care service category
HUD	U.S. Department of Housing and Urban Development

LM	loosely managed
MEF	Master Enrollment File
NRC	National Research Council
OIF	Operation Iraqi Freedom
OEF	Operation Enduring Freedom
OMB	Office of Management and Budget
RVU	relative value unit
SOE	Survey of Enrollees
UCPM	Unit Cost Projection Model
UPM	Utilization Projection Model
VA	U.S. Department of Veterans Affairs
VetPop	Veteran Population Model
VHA	Veterans Health Administration
VERA	Veterans Equitable Resource Allocation
VISN	Veterans Integrated Service Network
WM	well managed

CHAPTER ONE

Introduction and Background

The Department of Veterans Affairs (VA) uses a complex model—the Enrollee Health Care Projection Model (EHCPM)—as a planning tool to estimate demand for medical services and related support among U.S. veterans. The utility of the EHCPM depends on the model's ability to provide accurate and timely projections of future demands on VA resources consistent with the VA's budget and strategic-planning objectives. In order to understand the model's utility, the VA asked RAND, in conjunction with an independent, senior-level actuary (James P. Galasso, FSA, MAAA, President of Actuarial Services & Financial Modeling, Inc. [ActMod]), to conduct a comprehensive assessment of the EHCPM. This assessment included two specific sets of tasks: (1) a review and evaluation of key model features and (2) an assessment of the benefits and costs associated with the current specification and several specific aspects of the current contractual arrangement. This monograph presents the results of that assessment.

Overview of the Current VA Health Care System

The VA operates the largest integrated health care system in the United States. In 2005, the VA had 7.2 million enrollees, served 4.8 million unique patients, and had a total operating budget of $30.8 billion. Currently, all veterans with at least 24 months of continuous active-duty military service and an "other-than-dishonorable" discharge are eligible to enroll in the VA.

The VA provides primary and specialty care to enrollees, as well as a comprehensive pharmaceutical benefits program and other ancillary services. There are currently 1,264 VA health care facilities, including inpatient medical centers, outpatient clinics, and "Vet Centers" that provide assistance to the homeless and specialized counseling. The VA maintains partnerships with numerous academic medical centers to enhance quality of care and to promote education and training. In addition, the VA has developed an award-winning electronic medical records system and is considered an innovator in the field of health information technology. Recent studies have shown that veterans receive better overall quality of care than the general population (Asch et al., 2004; Etzioni et al., 2006; Stineman et al., 2001). For example, VA rates for colorectal cancer screening are higher than the national average (Etzioni et al., 2006), and VA stroke patients have better functional outcomes than comparable non-VA patients (Stineman et al., 2001).

The VA is one of the country's purest examples of a "staff model" health care delivery system. Specifically, the vast majority of the services provided by the VA are delivered in facilities owned and maintained by the VA and staffed by VA employees. The remaining services, referred to as "purchased services," are paid through negotiated fees and contracts. The VA provides a number of specialized services to address the unique needs of military veterans, including treatment for blindness, spinal cord injury, traumatic brain injury, mental illness, and post-traumatic stress disorder.

Copayments vary by the veteran's priority level. Veterans who meet certain conditions, such as having a qualifying, service-connected disability or a low income, receive care for free. Copayment rates for inpatient and ambulatory care services for upper-income veterans without service-connected disabilities are comparable to those required by Medicare. At the same time, copayments for prescription drugs are generally lower than those required by private health insurance plans and Medicare. Currently, the VA does not require an enrollment fee and requires no deductibles.

Because VA enrollment and use of VA services once enrolled are both voluntary, veterans who use the VA do so because they perceive it to be their best available option. Veterans for whom the VA is the best treatment option are likely to be those without other sources of health insurance coverage. As a result, veterans who rely heavily on the VA for their care tend to be less healthy and less well-off financially than the general civilian population (Ahga et al., 2000). The health needs of VA patients with service-connected disabilities may be especially different from those of the civilian population, given the effect of disabilities on health status and on access to medical care as a result of reduced employment opportunities (Angrist, 1990; Cohany, 1992; Savoca and Rosenheck, 2000).

Eligibility Reform

While the VA's historic mission is to provide a health care safety net for disabled veterans, the Veterans' Health Care Eligibility Reform Act of 1996 (PL 104-262) expanded the types of services available to VA patients and extended coverage to all veterans through a priority-based enrollment system. Care through the VA was historically available only to veterans who qualified under a complex set of income and disability requirements that were applied differently for inpatient, outpatient, and long-term care (Iglehart, 1996). Effective in fiscal year (FY) 1999, veterans were prioritized for enrollment according to eight tiers: those with service-connected disabilities (priority levels 1 and 2); prisoners of war and recipients of the Purple Heart (priority 3); veterans with catastrophic disabilities unrelated to service (priority 4); low-income veterans (priority 5); veterans who meet specific criteria, such as having served in the 1991 Persian Gulf War (priority 6); and all other veterans (priorities 7c and 8c). Appendix B provides detailed priority-level definitions. The change in the system was accompanied by large increases in enrollment; between 1999 and 2004, enrollment grew from 4.2 million veterans to 7.4 million veterans (Congressional Budget Office, 2005), an increase of 78 percent. Prior to the implementation of the Eligibility Reform Act, the VA was organized around medical centers that provided inpatient care. If a patient needed an outpatient service, he or she would have to have an inpatient admission in order to receive a referral. After eligibility reform, the VA began providing comprehensive care with a focus on outpatient and ambulatory care, effectively establishing a uniform medical-benefits package for all enrollees.

A key provision of the Eligibility Reform Act was the introduction of Community-Based Outpatient Clinics (CBOCs) located in areas that are far from a medical center and that have a relatively high concentration of veterans. CBOCs have improved veterans' access to care and have been a source of preventive care that can potentially alleviate conditions before they require more-specialized and expensive care.

When implementing the Eligibility Reform Act in 1996, the VA instituted a new organization centered on Veterans Integrated Service Networks (VISNs). There are currently 21 VISNs in the country, and each manages the budget for the medical centers and other service providers in its designated region. The VA's annual congressional appropriation is divided among the VISNs through the Veterans Equitable Resource Allocation (VERA) system. Instituted in 1997, the VERA system allocates funds based primarily on the number of veterans served.

Veterans are not entitled to VA health benefits by statute. Instead, the VA system relies on a discretionary budget. To assure that funding is adequate to meet the health care needs of its enrollees, the Secretary of Veterans Affairs can increase cost-sharing provisions or suspend new enrollment. Currently, priority 7 veterans (those with incomes above the VA-established threshold but below the Department of Housing and Urban Development [HUD] geographic index) and priority 8 veterans (those with incomes above the HUD geographic index) are subject to cost sharing. In 2003, enrollment of priority 8 veterans was suspended, and it continues to be suspended as of this writing.

Budget and Strategic Planning Under Eligibility Reform

While eligibility reform substantially expanded veterans' access to the VA health care system, it complicated both the budget- and strategic-planning processes. Prior to eligibility reform, facility capacity dictated the number of patients served by the VA health system. Thus, the VA's budget and strategic-planning processes considered the cost of staffing, maintaining, and expanding its facilities but not projections of patient demand. The Eligibility Reform Act required the VA to meet the full range of health care needs of an enrolled population in the absence of a statutory entitlement. Its mission under the Eligibility Reform Act requires the VA to budget and plan more like an integrated delivery system and less like a hospital system.[1] As a result, accurate forecasts of future resource requirements are vital to ensure that the VA's budget request is sufficient to provide comprehensive, integrated care to all eligible veterans. Accurate forecasts are particularly important because the VA has a limited suite of policy levers that it can use to reduce demand in the event of a budget shortfall. These levers include restricting future enrollment, increasing cost-sharing requirements for certain priority groups, and controlling the timing of capacity expansions and the level at which capacity expansions are funded.

Opportunities and Challenges of Policy Models

Models can provide policymakers with detailed, quantitative support for policy decisions, budget formulation, and strategic planning and management. In essence, a model is a simplified representation of reality that can provide insights into complex relationships (Stokey and Zeckhauser, 1978; Quade, 1989). Models range in complexity from a basic flow chart to an intricate computer simulation package. Models offer other benefits to their sponsoring organizations. Because models draw on a wide range of mission-relevant data sources and expertise to inform issues of strategic importance to their sponsoring organizations, participation in the development and maintenance of models can enhance organizational knowledge and learning. Policy models typically address topics central to an organization's operating mission and rely heavily on electronic data resources. As a result, models offer the opportunity to identify deficiencies in information technology systems and to target improvements to areas of strategic importance. While

[1] We use the term *integrated delivery system* to refer to an entity that employs, owns, and operates a substantial portion of the resources required to deliver health care to a defined population of covered lives.

models can provide a quantitative rationale for budget and planning initiatives, they can also become objects of intense scrutiny (Remler, Zivin, and Glied, 2004). The credibility of model-supported policy initiatives rests on the perceived model quality and objectivity. Yet the quality of complex models is difficult to demonstrate, particularly to nontechnical audiences. When quality is difficult to demonstrate, outsiders may look for other indications of quality, such as whether the model has been reviewed by independent experts and the professional credentials of model developers.

Actuarial Models

Actuarial models can help organizations that face uncertain demand for health benefits (such as the VA) to rationalize budget- and strategic-planning by quantifying financial liabilities under alternative assumptions about the future. Although there is no standard definition of an actuarial model (or, more specifically, an actuarial projection model), three defining characteristics of actuarial projection models are their

1. use of mathematical models to project future financial outcomes under uncertainty
2. heavy reliance on client-specific historical experience and administrative data sources
3. minimal use of explicit assumptions regarding the behavior of patients and clinicians.[2]

Limitations of VA Data for Forecasting Demand

Accurate forecasting requires comprehensive information about the types, amount, and cost of care that enrollees are likely to use. Our discussions with VA leadership suggest that the utility of VA administrative data for the purpose of demand forecasting is hindered by shortcomings in accuracy, consistency, and completeness.[3] However, we did not undertake a review of VA data systems sufficient to enable

---

[2] See Part 5 in Booth, 1999, for a discussion of actuarial models.

[3] These limitations characterize administrative health and utilization data generally and are not unique to the VA.

us to assess whether such coding deficiencies are more severe than those faced by other health benefit payers and providers (e.g., Medicare, large employers, Kaiser Permanente). At the same time, several unique features of the VA under the Eligibility Reform Act complicate the use of actuarial models to forecast future demand for VA services.

First, reliance on the VA is difficult to measure for younger enrollees who are not Medicare beneficiaries. Although most veterans are required to enroll with the VA in order to access VA services, there is no requirement that enrollees use the VA for any or all of their care. As a result, utilization documented in VA data systems represents a fraction of the total health care services used by enrollees. By linking VA and Medicare administrative data, the VA has a fairly complete understanding of the degree to which enrollees covered by Medicare rely on the VA for their care. However, the same is not true of enrollees not covered by Medicare. For example, VA administrative data documents roughly 30 percent of acute hospitalizations and 38 percent of outpatient care used by this younger group, and there is no central and readily available data source documenting non-VA care.

Second, the capacity of VA facilities to treat enrolled veterans varies geographically. Depending on the region, one can assume that at any given moment, demand for VA health care services exceeds, meets, or falls short of the VA's capacity to meet the corresponding demand. In regions operating under capacity constraints, VA utilization data will understate demand for VA services.

Third, issues related to reliance and capacity notwithstanding, substantial changes in utilization patterns following eligibility reform mean that historic data have only limited utility for informing future enrollment, utilization, and cost trends.

Because historical VA data reflect capacity constraints and provide only minimal information about reliance, simple models that rely on trends in such data are likely to perpetuate historical mismatches between demand for VA care and treatment capacity. Moreover, historic VA data may provide an insufficient basis for projecting changes in reliance and consequent changes in utilization and cost. Accurate forecasts require information about demand for VA services unconstrained by VA capacity at a given level of reliance.

The EHCPM: Description, History, and Uses

The EHCPM uses actuarial methods to project total expenditures in any given year by combining output from three model subcomponents—enrollment level, utilization rate, and unit cost—in a multiplicative fashion separately for each of 58 services for roughly 40,000 enrollee types (or "cells") defined by age category, whether enrollment occurred pre– or post–enrollment reform, priority level, and geographic sector. A key feature of the model is its use of proprietary utilization benchmarks based on the experience of commercial health insurers as a means of overcoming the limitations of VA data described above. The model applies four types of trend factors to account for general changes in medical costs and the anticipated changes in the efficiency of VA providers: utilization, inflation, intensity of service provision, and what Milliman refers to as the "degree of community management (DoCM)." These trends are in addition to the model's accommodation of assumed changes in veteran morbidity and veteran reliance on the VA health care delivery system, enrollment levels, and enrollment mix.

The model projects expenditures for a subset of services offered by the VA health system. Expenditures for modeled services comprise roughly 83 percent of the budget of the Veterans Health Administration (VHA). Modeled services include "Special VA Program Services," such as mental health case management, residential treatment for post-traumatic stress disorders and substance use disorders, blind rehabilitation, and methadone treatment. Nonmodeled services, such as long-term care, are modeled separately by the VHA and were not the focus of this project. Also, depreciation and the cost of capital attributable to facility construction and major renovations are not considered in the EHCPM. Thus, the model projects only what can be considered operating expenditures. Because the VHA budget is planned approximately three years in advance, the EHCPM is designed to develop budget estimates that reflect anticipated changes in enrollment. In addition, the EHCPM incorporates a wide range of service- and source-specific trend assumptions that are not accounted for in methodologies that inflate historic expenditures using a single trend factor. The model is described in more detail in Chapter Three.

The EHCPM has evolved over time and serves a variety of functions. The model originated in 1998 to support the Secretary of Veterans Affairs's annual enrollment decision as mandated by the Veteran's Health Care Eligibility Reform Act of 1996, which required the VA to establish a priority-based system for managing access to a fixed level of available resources. In its initial form, the model projected a single fiscal year into the future. In 2000, the model was modified to support the VA's Capital Asset for Realignment Enhanced Services (CARES) initiative, in which 20-year projections were used to estimate future demand for VA health care across geographic regions.

The model is currently used to develop the VHA's annual budget, the planning for which begins two and a half to three years prior to the final appropriation. The VA would like to extend the model's use to support policy analysis and strategic planning at the national, regional (VISN), and facility levels. Intended applications specifically include tests of the sensitivity of utilization and expenditures to cost-sharing requirements, the identification of unmet demand at the local level, and the sensitivity of demand to expanded access. See Appendix C for a more detailed list of intended uses as described in training materials developed by Milliman.

CHAPTER TWO

Evaluation Methodology

We undertook a series of activities in order to become familiar with the structure, accuracy, and utility of the EHCPM and to assess the model's overall benefits and risks to the VA. In carrying out these activities, we developed and were guided by a framework based on the relevant modeling literature. In this section, we describe our activities and framework, concluding with a description of the major limitations of our evaluation.

Key Activities

Established evaluation criteria. We identified and adapted evaluation criteria based on a review of actuarial standards and the literature concerning the specification and evaluation of large, complex policy models.

Conducted a detailed review of model documentation and supplementary materials. We carefully reviewed over 700 pages of documentation describing the base year (2002) version of the model and training materials prepared by VHA's Office of the Assistant Deputy Under Secretary for Health for Policy and Planning and Milliman staff for the purpose of familiarizing VA leadership and external constituents with the model's purpose, capabilities, and structure.

Participated in four training sessions. We held a 2-day session at RAND's Arlington, Va., office at which Milliman staff provided a broad overview of model components and supporting data. We attended a second 2-day session held at Milliman headquarters in Seattle, Wash., where Milliman staff engaged in detailed discussion on topics selected by RAND, ActMod, and VA staff, and provided an overview of programming logic used to produce model projections and standardized reporting of model results. This meeting also provided an opportunity to view model components propriety to Milliman. We attended a half-day meeting at VA headquarters in Washington, D.C., where Milliman staff reviewed for us the methodology for formulating model adjustments that account for regional variation in clinical management processes and the process used to formulate trend assumptions employed by the model. Finally, we attended a half-day meeting at VA headquarters at which VA staff from the Office of the Chief Financial Officer and the VA's Allocation Resource Center provided an overview of the VA's cost accounting system.

Reviewed validity studies and sensitivity analyses. We reviewed the results of four reports prepared by Milliman to assess the accuracy of model projections. The first report provided detailed comparisons of projected utilization in FY 2003 to actual utilization in FY 2003 by service category, enrollee type, and VISN. The second report compared 1-, 2-, and 3-year expenditure projections to actual VA budget obligations for services accounted for in the model. The third report compared the projected impact of a $5 increase in the prescription drug copayment rate on the use of prescription drugs for priority-level 7c and 8c enrollees to the actual change in prescription drug use following the copayment increase. The fourth report was a sensitivity analysis prepared by Milliman in response to a request made to the VA by the White House Office of Management and Budget.

Obtained information from VA and Milliman on an ad hoc basis. Throughout our evaluation, we contacted VA and Milliman staff on an ad hoc basis numerous times requesting additional information and clarification on a range of topics, including model features, staffing and level of effort specified in the VA's contract with Milliman, and the participation of VA staff in model development.

Developed a small prototype of the EHCPM. In order to test our understanding of the model mechanics and to gain insight into how the methodology allows assumptions about the VA's cost structure to influence expenditure projections, we developed a small prototype of the model. Our prototype was based on a sample calculation provided to us by Milliman staff that illustrated the methodology used to project future expenditures for office visits for a single cell of the model. This illustration is discussed in Chapter Four of this report.

Criteria for Evaluating Policy Models

There are no universally established standards for assuring the quality of large, complex models, such as the EHCPM, that are developed to inform policy decisions. While it may be intuitively appealing to use accuracy as the primary criterion for evaluating model quality, assessing model accuracy can be very challenging. Because such models are often used for long-term forecasting or to assess policy scenarios that may never come to pass, models cannot be assessed simply by looking at short-run projections and comparing them to realized values. As we discuss in Chapter Three, the EHCPM has several features that make standard approaches to assessing the accuracy of model projections potentially misleading. Thus, it is important to take a broad approach in assessing the quality of the EHCPM. In the sections below, we discuss criteria proposed by analysts and auditors for evaluating the quality of large, complex policy models (GAO, 1979; Stokey and Zechauser, 1978; Quade, 1989; National Research Council [NRC], 1991). These criteria include validity, accuracy, tractability, and transparency.

Validity

Validity refers to the ability of the model to fulfill its intended purpose. In other words, a model is valid if the estimates and projections derived from the model are consistent with the model sponsor's objectives in developing and maintaining the model. Importantly, a model can be valid for one set of purposes and not for others.

In order for a model to be valid, its structure should reflect relevant features of the policy environment and avoid nonessential elements. Models designed to support policy decisions must capture both features of the environment that influence the system being modeled and relationships between policy levers and relevant policy outcomes.

In the context of projecting demand for public health care programs, essential features might include the generosity of the public benefit relative to other alternatives available to program beneficiaries and the relative mix of health risks the benefit might attract. In the context of expenditure projections, essential features include whether the program delivers services directly, provides services through risk-based contracts with private organizations, or pays fee-for-service according to a predetermined fee schedule.

Published professional standards advise actuaries preparing financial projections to "consider the historical experience of the insurance business, adjusted to reflect known material changes in the environment and identifiable trends to the extent such information is available" (Actuarial Standards Board, 2005). In the context of projecting future veteran health care expenditures, we believe that this standard is consistent with the use of as many internal VA data sources as possible in model development. Emphasis on the use of internal data sources helps to minimize threats to validity resulting from unmeasured differences between client characteristics and the characteristics of the external data source. Quite often, however, actuaries must supplement a client's actual data with data from external sources. Two situations that necessitate the use of external data are (1) when the client's data are considered to be less than fully credible due to insufficient size, integrity, or stability and (2) when the parameters to be projected or used in the projection process are new to the client or not represented by the available client data. In general, the larger the client, the larger the available actual data for projection purposes and the least reliance need be placed on external sources.

Assumptions are conditions that must reasonably hold in order for a model to be valid. Assumptions can be explicit or implicit. Explicit assumptions are easy to observe based on a model's structure or parameter values. The EHCPM, for example, makes explicit assumptions about the relationship between utilization and the age and gender of enrollees. Implicit assumptions, by contrast, are more difficult to detect and assess because they arise as an indirect consequence of model structure.

If either type of assumption is not reasonably met, model results may be inaccurate or invite an invalid interpretation. Assumptions should be clearly stated, allowing both internal and external stakeholders to assess their credibility (GAO, 1979). Model developers should make stakeholders aware of key assumptions, demonstrate their reasonableness, and where possible use sensitivity analyses to assess the influence of key assumptions on model results. The influence of explicit assumptions on model results should be assessed through sensitivity analysis, which involves replacing model inputs (such as parameters) with reasonable alternatives and re-estimating. The resulting output can then be compared against a baseline to gauge the model's sensitivity to particular assumptions. Sensitivity analyses that yield wide ranges may help to flag trouble spots, particularly if there is already an indication that the model is performing poorly. The reasonableness of implicit assumptions is more difficult to demonstrate. In many cases, implicit assumptions can often be supported on the basis of reason, expert judgment, or evidence from the research literature. While not always practical to implement, the sensitivity of model projections to implicit assumptions can be demonstrated by comparing projections from alternative model specifications that differ in their reliance on the implicit assumption in question.

Accuracy

Model validity contributes to the accuracy of model forecasts. At the same time, model accuracy is not sufficient for validity because models can be correct by accident or be structured in such a way that they do not inform key policy issues facing the sponsors. From the perspective of model constituents, accuracy is probably the most salient and comprehensible feature of any complex policy model. Yet, accuracy can be surprisingly difficult to assess. Ideally, the accuracy of forecasting models can be assessed by comparing the forecasted outcome with the known outcomes (once they have been realized). Based on the known outcomes, modifications to the model structure aimed at improving accuracy can be made and accuracy reassessed. At the same time, however, uncertainty about model accuracy is an inherent concern when using forecasting models to inform policy decisions. This is because the accuracy of the forecast is not known when the decision must be made. The accuracy of "what if" scenarios is often never known because the "what ifs" are rarely implemented, meaning that the information needed to assess accuracy is never generated. Assessing the accuracy of projections used in VA budget planning poses special challenges because approved budgets reflect the outcomes of political and strategic decisions—the outcomes of which projection models are not intended to predict.

Model developers should establish a framework and predetermined criteria for assessing model accuracy. In the case of policy models such as the EHCPM, however, the development of a comprehensive approach to assessing model accuracy that explicitly relates the accuracy of model subcomponents and assumptions to overall forecasting error may be too complex and time consuming to be practical. Instead, model developers should compare projections to realized outcomes when possible. In situations where it is difficult to assess accuracy in this way, model developers should help model sponsors to understand and communicate these limitations to model constituents and propose specialized analytic approaches for demonstrating accuracy. In the case of budget-planning applications, for example, special procedures should ideally involve the creation of a set of revised projections tied to the resource levels consistent with the approved budget.

Even if model theory and assumptions are sound, estimates will be inaccurate if computer programs underlying the model are flawed or mathematical algorithms used to compute projections are incorrect. In order for the computer model to be accurate, it must incorporate all relationships specified in the theoretic model, and it must be properly mechanized so that it runs as expected (GAO, 1979). Assessing the quality of computer code requires knowledge of the languages used, access to all programs involved, and—ideally—the ability to run each step of the program to check output for errors. Model developers can facilitate this process by maintaining detailed records of all programs involved in model execution, annotating programs, and keeping a log of program updates and corrections. While this documentation process can be time-consuming, it helps to ensure that future programmers can easily understand and execute the model code.

Model developers should provide model sponsors with a sense of the certainty surrounding model estimates. For some types of models, it is possible to estimate the variance around point estimates using well-established statistical techniques. Such techniques often do not exist for very complex models, although in some cases they can be specially developed (e.g., bootstrap procedures) (NRC, 1991). When model certainty cannot be practically assessed using statistical methods, model developers can provide a sense of the degree of certainty through a range of estimates, based on judgments about the likely range of parameters, that describe key features of the policy environment.

Tractability

Tractability refers to the ability of model developers, sponsors, and constituents to understand the model's structure and operations. To assure tractability, model developers must strike a balance between essential features and those that can be disregarded for the sake of clarity. At first blush, it might seem that the most effective models would strive to replicate real world situations by incorporating as much detail and complexity as possible. While increasing model complexity can increase validity by making a model more realistic, overly complex models are difficult for stakeholders to understand and can exceed the modeler's ability to document, update, assess, and replicate key model features (GAO, 1979; Stokey and Zeckhauser, 1978; Quade, 1989; NRC, 1991). Moreover, a model structure that cannot be measured and monitored with readily available data reduces the model's practical relevance.

Component-based models combine input from several self-contained modules to produce final output. Because they allow the user to change each particular sub-component without rebuilding the entire model, component-based models can substantially increase tractability. Component-based models make it easier to assess or modify model outcomes that involve only a single component (NRC, 1991). Examples from the EHCPM include enrollment rates, enrollee type, and utilization rates for a particular type of service. Component-based models also reduce model complexity by using the outputs of other modeling efforts as model inputs in an orderly fashion. For example, the EHCPM uses projections of the future veteran population from the VetPop model sponsored by the VA Office of the Actuary as a key input in the projection of future health system enrollment. NRC (1991) recommends that each module within a component-based model should itself be "highly parameterized." That is, the user should be able to change various features of the model, such as inflation trends or demand elasticities, by substituting alternative parameters rather than rewriting computer code. A highly parameterized, component-based structure ensures that the model can easily be adapted to respond to emerging policy questions.

The use of expert judgment to establish parameter values can reduce tractability. When existing data sources are not sufficiently detailed, model developers face two choices: (1) use expert judgment to "fill in" or "parameterize" the missing data or (2) modify/simplify the model structure to match the available data detail. Because expert judgment often cannot be readily validated or verified, model developers should employ expert judgment judiciously.

Transparency

Model developers should strive to make the model structure and modeling process as transparent as possible. Lack of transparency can undermine a model's credibility, particularly in the context of a public-sector model where key constituent groups do not have institutional understanding of the model sponsor's organization. The need for transparency is underscored by the facts that complex policy models are rarely subject to peer review and there are no universally established standards for reporting the assumptions used by the model (Glied, Remler, and Zivin, 2002). Because models are often highly technical, maintaining transparency can be quite challenging.

Comprehensive and comprehensible documentation contributes significantly to ensuring that the model is transparent, that results can be replicated by interested parties, and that personnel changes and gaps or transitions in sponsorship will not disrupt model continuity (GAO, 1979; Gass and Thompson, 1980; NRC, 1991). There are two types of documentation described by the General Accounting Office (GAO, 1979). Descriptive documentation consists of general information about the model, including its theory, assumptions, and limitations. The purpose of descriptive documentation is to provide information to constituents who want a high-level understanding of the model's uses and capabilities. In contrast, technical documentation should be sufficient to permit outside parties to operate the model, validate the model's results, or replicate the model's findings.

In addition to containing relevant content, model documentation must communicate clearly to its intended audience. In a case study that assessed documentation for a sample of complex policy models, NRC (1991) found that the quality of model documentation was frequently undermined by seemingly minor copyediting issues. Typical problems included typographical errors, the failure to describe acronyms, inconsistent formatting, jargon, poor writing, and inaccuracies. NRC also found that model versions were often ambiguous in the documentation, leading to obsolete cross-references and outdated information. Finally, NRC found that the sheer size of model documentation could make it challenging to understand, particularly in the absence of a comprehensive index.

Although good documentation is essential to ensuring the longevity and credibility of a model, high short-run costs may cause reluctance in investing in documentation (GAO, 1979). Both model sponsors and developers must be aware of this issue and work to maintain documentation despite the fact that adequate recordkeeping can be arduous and time-consuming. NRC (1991) suggests that sponsors consider executing a separate contract for model documentation or explicitly requiring the skills of a technical writer for the purpose of documentation in the model development contract. Such arrangements would ensure that modelers do not push aside documentation efforts due to the more urgent pressures of model development.

Peer review is an additional avenue through which to improve model transparency. Encouraging periodic review by independent analysts can dispel concerns that the model is unduly influenced by internal stakeholders. Further, this type of review increases the probability that model flaws and oversights will be discovered and corrected. For example, actuarial models used to project future Social Security and Medicare payments are reviewed by panels of experts comprised of economists and actuaries (GAO, 2003; Old-Age, Survivors and Disability Insurance Trustees, 2006).

Limitations

Our engagement contract, resources, and time constraints limited the depth and scope of our evaluation. As a result, we did not consider in depth potentially relevant information from several sources. First, we did not review in depth VA data systems and surveys that serve as the basis for many model inputs. Second, we did not review in depth components proprietary to Milliman or the methodology Milliman uses to produce them. Third, we did not conduct a detailed review of computer code. Fourth, we did not explore in depth the feasibility and relative merits of alternative modeling approaches, including those currently under development. Fifth, we did not review the statement of work and budget stipulated in the VA's contract with Milliman. Finally, we did not investigate in depth the VA's capacity to bring the model or key modeling functions in house.

Limitations to the scope and depth of our evaluation shaped our findings and recommendations in two important respects. First, our limited review of VA data systems means that we were not able to assess fully the extent to which the model exploits the potential of VA administrative data sources. Second, because we did not identify specific alternatives to the current model and did not review detailed information about the cost of the EHCPM contract, we were not able to conduct a formal assessment of the costs and benefits of the model. Despite these limitations, we are confident that the scope and depth of our evaluation was sufficiently comprehensive and informative to permit us to formulate findings and recommendations that will be meaningful and useful to the VA.

CHAPTER THREE

Model Overview

To guide their application of the assessment criteria, the team first set out to understand the EHCPM's basic structure and operations. This chapter reports the results of this descriptive analysis. The EHCPM is a large and complex model that combines data from a wide range of sources to project the future costs of providing health care services to enrolled veterans. The model projects future annual enrollment, utilization, and operational expenditures for a comprehensive array of health care services by key enrollee characteristics over a 20-year period.

Milliman Health Cost Guidelines

The model is structured around a proprietary set of utilization rates based on private-sector data and adjusted to reflect demand by an enrolled veteran population for a comprehensive set of services offered by the VA. The utilization rates are a component of the Milliman Health Cost Guidelines (HCGs), which are a pricing and utilization benchmarking tool commonly referred to as a *rating manual.* Milliman's HCGs have been marketed for more than 50 years, and they are currently used under license by a large clientele of actuaries, health plans, and employers throughout the United States. The HCGs provide utilization rates and expected claim costs for 37 of the 58 EHCPM health care service categories. The HCG utilization and cost data are based on a standard benefit package for a wide range of demographic characteristics and geographic regions. The HCGs also contain information regarding the responsiveness, or the *elasticity*, of utilization and claim costs to changes in benefit design.

The data underlying the HCGs come from a variety of sources, some of which are familiar to actuaries, health services researchers, health economists, and clinical epidemiologists and include

- Medicare claims data
- purchases of commercially available claims databases, such as Medstat MarketScan
- administrative data from individual health plans obtained through data trades, research agreements, and consulting engagements
- publicly available hospital discharge data from the Agency for Healthcare Research and Quality's Healthcare Cost and Utilization Project.

The HCGs are developed through a combination of statistical analysis, actuarial modeling, and the expert judgments of actuaries. The EHCPM is similar to other applications for which Milliman actuaries combine the HCGs and client-specific data using a standard set of modeling approaches. At the same time, the EHCPM application is unique, given the size of the VA system, in its degree of reliance on the HCGs as opposed to client-specific, historical experience. Discussions with VA and Milliman staff suggested that changes in the structure and size of the VA accompanying eligibility reform and in the organization and delivery of health care services limit the utility of VA administrative data for modeling purposes.

The intended function of the HCGs in the EHCPM is to provide information about the effect on utilization rates of changes over time in the demographic composition of the veteran population and of changes in efficiency of VA facilities relative to the communities in which they operate. The HCGs are also intended to compensate for gaps in VA administrative data that reduce their suitability for projecting future demand for services resulting from constraints on the availability of VA services and enrollees' partial reliance on the VA. The HCGs provide estimates about the sensitivity of demand to changes in benefit design. Because the VA benefit has remained stable in the period following enrollment reform, external information is required to estimate the responsiveness of demand to any future changes in copays and deductibles. Finally, the HCGs are intended to provide a source of reliable, or *credible*, utilization rates for services for which VA data may be lacking. These services include those that are used rarely (e.g., obstetrics), that have historically been supplied in limited amounts (e.g., chiropractic), or for which service definitions or internal VA coding conventions may have changed over time (e.g., physical therapy services).

Overview of Model Structure

The EHCPM is based on standard actuarial formulas and modeling techniques. The high-level structure of the EHCPM is illustrated in Figure 3.1, in which enrollment, utilization rates, and costs are multiplied together to produce expenditures in any given projection year. Inputs to this equation are derived from three separate submodels that are assumed not to interact: (1) the Enrollment Projection Model (EPM), (2) the Utilization Projection Model (UPM), and (3) the Unit Cost Projection Model (UCPM).

Expenditures are obtained by multiplying the projected units of service by the projected unit cost. Total projected expenditures in a given year are obtained by summing across cells and service categories. Each of the model subcomponents and the trending of modeled utilization and unit costs are described in greater detail below.

While the relationships among the submodels are relatively simple, the EHCPM's complexity stems from a series of adjustments and imputations to the submodel components required to derive enrollment, utilization rates, and average unit costs at a fine level of service detail that reflect VA case-mix and care intensity. Adjustments are derived from VA data analyses, external data sources, statistical modeling, and the informed judgments of Milliman actuaries and senior VA leadership and program staff.

Figure 3.1
High-Level Structure of the EHCPM

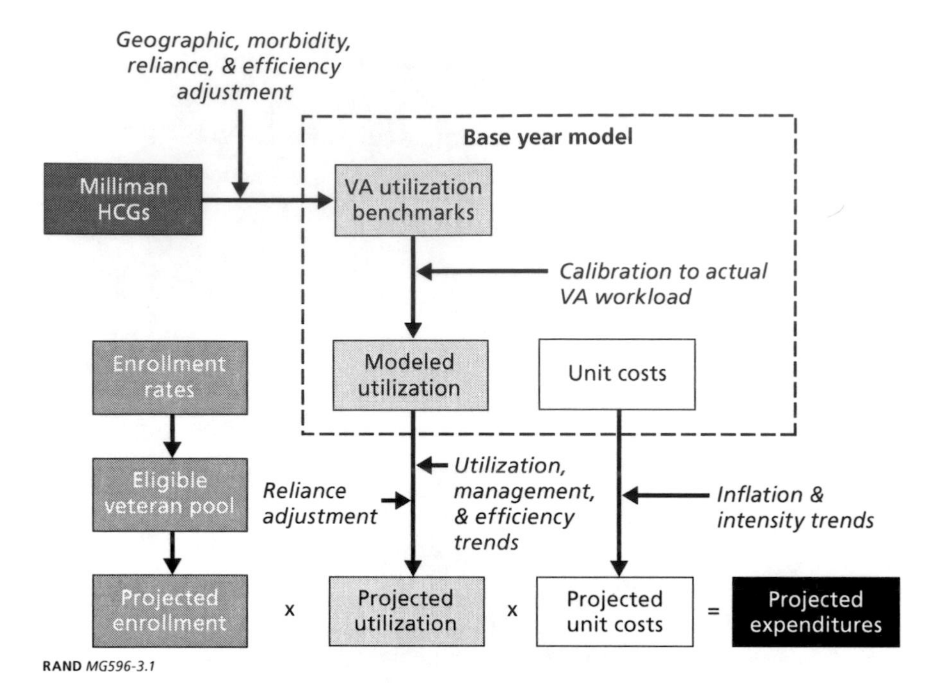

RAND *MG596-3.1*

Enrollment Projection Model

Inputs for the enrollment model are derived from projections of the veteran enrollee population and historic enrollment rates. The EPM projects the number of enrollees in a given cell defined by geographic sector, age category, gender, enrollee type, and priority level. In any given projection year, the EPM divides the veteran population into two groups: (1) enrolled veterans and (2) those in the non-enrolled pool. Enrollment in any given cell in any projection year is equal to current enrollment plus new enrollment minus deaths. New enrollment is obtained by applying enrollment rate estimates derived from historical experience to the pool of non-enrollees.

Non-Enrolled Pool. In order to identify potential enrollees in the pool, the model draws from multiple data sources. Two of the key data sources are two 20-year projections developed by the VA Office of the

Actuary: (1) the Veteran Population Model (VetPop), which forecasts the veteran population by age, gender, and disability status using data from the 2000 U.S. Census and (2) the County Census 2000 VetPop Projection, which forecasts the population by 5-year age band, gender, and county of residence. Since priority level is not projected in either of the VetPop files, priority levels must be imputed using a variety of additional data sources. By combining the national-level information on disability from the VetPop with county-level variation in disability rates observed in the September 2002 Compensation and Pension file,[1] Milliman develops county-specific allocations to assign veterans to priority levels 1, 2, and 3. Additional information on national disability rates from the 1997 Survey of Income and Program Participation, as well as data from the VA, is used to assign veterans to priority 4 (catastrophically disabled but not due to a service-related injury). The proportion of low-income veterans (priority 5) is estimated using data from the decennial Census long form. Finally, the remaining veterans are attributed to priority levels 6, 7, and 8 as a combined pool for the purpose of estimating new enrollment.

Current Enrollees. The Master Enrollment File (MEF) is supplied by the VHA Office of the Assistant Deputy Under Secretary for Health for Policy and Planning. Data from the MEF provide the following information about enrolled veterans: (1) scrambled social security number, (2) priority level, (3) date of birth, (4) gender, (5) date of death, (6) enrollee type, pre–/post–enrollment reform, (7) county of residence, (8) preferred facility, (9) Operation Iraqi Freedom (OIF)/ Operation Enduring Freedom (OEF) first and last theater dates, (10) military retirement status, and (11) enrollment date. VA workload data describing first and last encounter dates are linked to the MEF. These data are used to assess and correct enrollment and death data and to support projections of the future number of enrollees and patients in VA facilities.

Enrollment Rates. The enrollment rate is the ratio of new enrollees to non-enrolled veterans in the pool. Historical monthly enrollment

---

1 Versions of the model more current than the one we were asked to evaluate use the 2004 VetPop file.

rates for the period between 1998 and 2005 are contained in the MEF and are extrapolated to obtain monthly enrollment rates by cell, where cells are defined using three age bands (<45, 45–64, 65+), priority level, geographic sector, and OIF/OEF status. When the number of enrollees is very small, sector-specific enrollment rates are extrapolated from more-aggregated geographic areas. Enrollment rates are assumed to be constant over the projection period based on an analysis of changes in historical enrollment rates. However, enrollment rates are adjusted downward in markets that are near saturation so that projected enrollment does not exceed the number of eligible veterans in a geographic area.

Monthly enrollment levels are projected for each cell. Enrollment is equal to current enrollment plus new enrollment minus deaths. New enrollment is equal to the product of the enrollment rate times the number of non-enrolled veterans in the pool. Enrollment estimates are also adjusted to reflect estimated changes in priority level and expected geographic migration among enrollees over time. Newer versions of the model adjust projections to reflect more-recent enrollment data, which are reported to Milliman on a monthly basis.

Utilization Projection Model

For the 37 (out of 58) health care services that are typically covered by private-sector health plans, projected utilization within the EHCPM is determined using commercial benchmarks from the Milliman HCGs that are calibrated to reflect the VA population. The UPM produces a trended utilization rate per 1,000 enrollees for each cell in each projected fiscal year for each of the modeled health care service categories (HSCs). For each population cell, HSC, and year, the projected utilization rate is multiplied by the projected number of average veteran enrollees in each year to produce the required units of service. Projected utilization for the remaining 18 services is modeled using an alternative methodology.

The UPM starts with national utilization rates from the HCGs and applies a series of adjustments that result in a VA-specific utilization rate for each cell. In the paragraphs below, we discuss the adjustments, their function, and underlying assumptions in general terms.

The series of adjustments that translates Milliman HCG benchmarks to VA-specific utilization rates is shown in Table 3.1, which was adapted from model training materials. This example shows the calculations underlying the projected office-visit utilization rate in FY 2008 for the cell specified in the first rows of the table, which is comprised of males age 55–59 who reside in the geographic sector labeled 03-c-9-F and who have priority-level 2 status. Utilization rates for other VA covered services with private-sector counterparts are derived through an analogous process. Services not typically covered by private-sector health plans, referred to as *Special VA Programs,* are modeled using an alternative set of procedures and are not the focus of this evaluation.

A key feature of the UPM is its potential capability to project the effect on service-specific utilization rates of differences in the efficiency of clinical management in the VA compared to the local community (sector) in which the modeled VA care is provided—the DoCM. The DoCM is imbedded in the model through a set of adjustments that place the VA utilization rate along a sector-specific utilization continuum. At the upper extreme, this continuum is bounded by the national average utilization rate experienced by a set of "well managed" (WM) health plans (considered 100-percent managed within the model) and at the lower extreme by the community average utilization rate for a set of "loosely managed" (LM) plans (considered 0-percent managed within the model). The WM designation is given to a proprietary set of health plan data meeting Milliman's standards for best clinical and utilization practices, and the WM benchmark is assumed to be constant regardless of geographic location. Plans that reimburse providers on a fee-for-service basis are used to define the LM bound. LM plans include certain preferred-provider networks and point-of-service plans, the identity of which Milliman treats as proprietary. Together, the local LM and the national WM utilization benchmarks create an interval, the length of which varies by local VA market area.

Geographic and Benefit Adjustments. The development of local efficiency intervals described above starts in line *a* of Table 3.1, with national LM and WM utilization rates corresponding to a standard, private-sector benefit package and a national case-mix. Line *b* adjusts the national LM benchmark to reflect local community practice pat-

Table 3.1
Utilization Projection Model Example

Label	Factor/Adjustment	Community Loosely Managed	National Well Managed
General			
a	HCG national average utilization rate per 1,000 (base year 2005)	2,988	2,587
b	HCG area adjustment	1.164	n/a
c	HCG copay adjustment	1.220	1.051
d	HCG covered-benefit adjustment	1.000	1.000
$e = a \cdot b \cdot c \cdot d$	Intermediate utilization rate per 1,000	4,243	2,719
VA specific			
f	VA DoCM score		22%
g (uses e and *f*)	Intermediate utilization rate per 1,000		3,907
h	HCG age/gender adjustment		1.282
i	VA reliance		0.512
j	VA morbidity		1.665
k	VA trend assumption		1.049
$l = g \cdot h \cdot i \cdot j \cdot k$	Modeled utilization rate per 1,000		4,479
m	Actual-to-expected adjustment (base year 2005)		0.929
$n = l \cdot m$	VA projected utilization rate		4,161
p	Projected enrollment		6
$o = p \cdot n/1{,}000$	Total units		25

SOURCE: Adapted by RAND and ActMod from training materials provided by Milliman, Inc., and the VA.

NOTE: Cell specifications are benefit—office visit; fiscal year—2008; geographic sector—03-c-9-F; enrollee type—priority level 2, pre (non OIF/OEF), males, age category 55–59, age cohort 50–54 in 2005.

terns. The geographic adjustment is not applied to the national WM benchmark under the assumption that regional differences in practice style do not influence utilization rates in WM plans.

In the next step (line *c*), the LM and WM utilization rates are adjusted to reflect differences in the utilization rate induced by differences in VA copayments compared to the standard benefit reflected in the HCG rates. The copayment adjustment strategy assumes that differences in copayments influence only the utilization rate and not the intensity of care. The WM copayment adjustment is smaller than the LM copayment adjustment under the assumption that utilization rates in WM plans are less sensitive to copayment levels. Elasticity assumptions underlying the copayment adjustments and the methodology used to derive them are proprietary and, thus, were not reviewed during our evaluation.

For a small number of services, a covered-benefit adjustment (line *d*) is applied to the LM and WM utilization rates to account for differences in utilization induced by differences in the VA benefit design at the HSC level compared to the standard benefit design used to develop the HCGs. Examples of services to which the covered-benefit adjustment is applied include ambulance and inpatient psychiatric stays.

The geographic and benefit adjustments described above result in HSC-specific utilization rates that national WM and local LM plans would be expected to experience were they to offer a benefit similar to the VA's and serve enrollees with a national case-mix (line *e*). The steps described below adjust this starting point to reflect VA management practices and the case-mix presented by the VA enrollee population.

DoCM. The next step establishes the placement of the local VA health care system relative to the interval formed by the LM and WM rates (line *e*). This VA-case-mix adjusted placement is measured as a percentage and is referred to as the *DoCM* (line *f*). For ambulatory care, a VA workgroup set DoCM rates between 6 percent and 20 percent, depending on the extent to which the local VA has implemented an initiative aimed at improving the efficiency of ambulatory care. Known as Advanced Clinic Access, this initiative was phased in during FY 2004–FY 2006. Projected DoCM levels were derived from implementation rates projected for the Advanced Clinic Access protocols. Assuming Advanced Clinic Access protocols are fully implemented, future increases in DoCM levels are expected to be 0.5 percent per year. For outpatient psychiatric and substance abuse, the model assumes a DoCM of 6 percent for 2004, increasing by 0.5 percent per year. In the case of pharmacy services, a VA workgroup that included senior pharmacy benefit-management staff estimated that the DoCM level for drugs and pharmacy services was at 40 percent and would increase 5 percent upon full implementation of Advanced Clinic Access initiative and 0.5 percent per year thereafter. In the current version of the model, a VA pharmacy workgroup has established DoCM at 90 percent in FY 2004. The pharmacy DoCM is assumed to increase 0.5 percent per year through FY 2007 and 0.25 percent per year thereafter until DoCM reaches 95 percent.

In the case of inpatient care, the DoCM levels result from longitudinal analyses using VA workload and Medicare administrative data from FY 2000–FY 2003 to compare the relative performance of VA facilities in reducing case-mix adjusted avoidable days. The validity of this calculation depends on the comparability of the case-mix adjustment methodology across VA and non-VA systems. The WM avoidable day rate is set at 100 percent, and the local LM rate is set at 0 percent. The VA DoCM level calculations can, and in many cases do, result in negative DoCM values in cases where the VA is less efficient than local LM plans. An internal VA workgroup set targets for improvement in DoCM at 2 percent of the way toward the WM benchmark per year, with facilities with a positive DoCM improving 0.5 percent per year.

In the current version of the model, targets for improvement in inpatient DoCM will be based on avoidable days (moving towards VA best practice). Each facility's DoCM was set at the measured level in FY 2005, and targets for improvement were based on historical improvement in VA avoidable inpatient days measured in 2001–2005. Facilities that achieve VA best practice will be assumed to improve at a rate of 0.5 percent per year until their DoCM level reaches 50 percent, after which DoCM will remain fixed. Special programs, ambulance, and durable-medical-equipment services do not have DoCM assumptions.

Once the cell-specific DoCM level is determined (line *f*), it is applied to the interval defined by the local WM and LM utilization rates. Applying the DoCM to the WM-LM interval (line *e*) results in a management level that the VA would be expected to achieve (line *g*). Because base year DoCM levels and improvements are specified in percentage-point terms and the width of the WM-LM interval varies from sector to sector (line *e*), the assumed changes in VA utilization rates varies from community to community as a function of the width of the local LM-WM interval.

Age-Gender Adjustments. Next, the UPM adjusts the utilization rate (line *g*) to be specific to the age/gender composition of the modeled cell. The validity of these adjustments depends on the assumption that differences in utilization by age and gender observed in the commercial sector are similar for veteran enrollees, after accounting for measured differences in health status addressed by the model's morbidity adjustment (described below).

Reliance. The local VA age/gender–adjusted utilization rate (lines *g* and *h*) is adjusted to reflect the estimates of the portion of all utilization demanded by VA enrollees that will be provided by the VA (line *i*). Calculations are different for over and under age 65. For the 65-and-over enrolled population, reliance can be observed by linking VA workload with Medicare utilization at the patient level. Reliance by service category by cell is estimated by dividing total VA utilization by total VA and Center for Medicare and Medicaid Services (CMS) utilization. Utilization is measured in days for inpatient services and billable current procedural terminology codes for outpatient visits, referred to in the VA as *clinic-stops*. An internal Milliman study suggests that, although CMS inpatient stays tend to be more complex than VA stays, accounting for differences in complexity would not have a large effect on calculated reliance rates. Reliance for the under-65 population is estimated using self-reported reliance in the 2002 and 2003 Survey of Enrollees (SOE) and adjusting for response bias using the observed relationship between reliance reported in the SOE by the 65-and-over population and reliance measured using linked VA-Medicare data. The under-65 reliance estimates assume that the service-level relationships between self-reported reliance and reliance measured with linked VA-Medicare data for those 65 and over hold for individuals under 65.

The model accounts for anticipated changes in reliance in projection years resulting from capacity expansions in selected markets for selected services. The model assumes capacity expansions will be absorbed fully and not result in excess supply. In the cases of cardiology, outpatient mental health and substance abuse, and VA special services, all markets were increased to a minimum level of reliance by moving all markets below the 85th percentile of reliance to the 85th percentile over various timeframes.

The model also adjusts for several markets with identified capacity constraints by increasing the estimated reliance levels. The calculation of the new reliance level involves a market-level regression of enrollee survey responses, VA performance on access measures, and enrollee travel distance on observed reliance on the VA, with markets considered by a VA workgroup to be capacity constrained removed from the model. This regression produces expected, or *target*, reliance, measured in percentiles. If the reliance percentile is less than the target percentile, then an increase in reliance is phased over a 3–4 year period. If the reliance percentile exceeds the target, it remains unchanged.

Morbidity. The local VA age/gender– and reliance-adjusted utilization rate (lines *g*, *h*, and *i*) is further adjusted to reflect differences in the health status of veterans relative to their counterparts in the general population (line *j*). Morbidity adjustments are calculated separately for enrollees over and enrollees under age 65 via a multistep process. For the enrollees 65 and older, the first step maps diagnostic codes to disease conditions using the Chronic Illness and Disability Payment System (CPDS), which is often referred to as the *CPDS grouper* algorithm (Kronick et al., 1998). Using Medicare claims data, indicator variables measuring the presence or absence of grouper conditions are then regressed to obtain estimated condition (or *morbidity*) weights on each condition by service category. Weights are used to calculate average expected Medicare expenditures by gender and age category. In the second step, the estimated condition weights are applied to linked VA and Medicare utilization data for the 65-and-over enrollees to obtain expected private-sector expenditure for all Medicare-eligible VA enrollees for each service category. These expected expenditures were averaged within age/gender cells by service category weighted by reliance. Reliance weighting accounts for the idea that only a fraction of additional utilization induced by higher morbidity will take place in VA facilities. Next, for each age/gender–service category cell, the ratio of expected private-sector spending for VA enrollees to private-sector spending for similar Medicare beneficiaries with a comparable age/gender mix is calculated, creating the morbidity adjustment shown in line *j*.

The lack of private-sector utilization data that can be linked to VA utilization data at the patient level complicates the derivation of a morbidity adjustment for under-65 enrollees. Derivation of the adjustment is complicated by potential confounding of morbidity and reliance in VA workload data. Milliman has been exploring alternative approaches for using the observed relationship between reliance and morbidity from both administrative and survey data among the 65-and-over enrollees to extrapolate expected private-sector expenditures for under-65 enrollees. These alternative methods for extrapolating under-65 morbidity from relativities in 65-and-over estimates (e.g., between morbidity and reliance) assume that 65-and-over relationships apply to the population under 65.

Actual-to-Expected Adjustment. The projected, or *modeled*, utilization rate (line *l*) is adjusted to reflect the ratio of the expected HCG-based utilization rate in the base year to the actual utilization rate in the base year, which is measured using VA workload data. The difference between actual and expected utilization represents a composite of factors that influence utilization but are not fully accounted for in the UPM. The actual-to-expected (A/E) adjustment compensates for unexplained variations in utilization in the base year and in projection years to the extent that model assumptions accurately account for future changes in the relative magnitude of model factors. Depending on the level at which the A/E and other utilization model adjustments are performed, model projections will be more or less influenced by the relative value of commercial utilization rates for different model cells as published in the HCG rating manual. Milliman refers to these levels or categories of adjustment as *A/E cells*. Projections made when A/E adjustments are performed at the aggregate national level, for example, will maximize the influence of the HCG "relativities." At the other extreme, projections made when A/E adjustments are performed at the cell level will minimize the influence of the HCGs. In the base year 2002 model, A/E adjustments are performed by priority level, enrollee type (pre/post), and broad age category (under 65 and 65 and older). In this way, A/E adjustments incorporate information about the degree of variation in VA utilization across enrollee categories not contained in HCG-based benchmarks.

Because A/E adjustments in the base year 2002 model do not vary across detailed age categories and because reliance and morbidity adjustments are not performed at the detailed age level, information about the relative influence of age on utilization contained in model projections comes solely from the HCGs. Likewise, information about the relative influence of geographic region on utilization comes from HCG area adjustments, DoCM adjustments, and geographic detail in reliance and morbidity adjustments. (Note that in the current version of model, A/E adjustments vary across geographic regions.) The A/E–adjusted utilization rate (line *n*) is multiplied by the projected number of enrollees in the modeled cell (from the EPM) and applied to the projected unit cost to develop expenditures. The derivation of unit costs is described in the subsequent section.

Special VA Programs. The utilization of services without private-sector counterparts is modeled using an alternative methodology that does not rely on private-sector benchmarks. The methodology we reviewed starts with a national base rate derived from FY 2002. However, the current version of the model uses FY 2003—FY 2005 workload data to establish the modeling factors. VA workload data are expressed in terms of utilization per 1,000 for each service category. The national base rate is adjusted by enrollee type, priority level, cohort, age/gender, and VISN.

Unit Cost Projection Model

The UCPM produces a trended average unit cost for each HSC by age/gender and market cell in each projection year. The VA's cost accounting system, the Decision Support System (DSS), provides the foundation for the base year unit costs used by the UCPM. Unit costs are derived by DSS through the allocation of the VA budget obligation for modeled services in the base year to actual VA workload for those services in the base year. This method of deriving unit costs does not relate the additional, or *marginal*, cost to the VA of producing an additional unit of service to treatment capacity or utilization levels. This costing method is similar to that used by fee-for-service health insurers such as Medicare, whose unit costs are essentially the prices charged by some external entity.

DSS-based unit costs can fluctuate substantially from year to year because unit costs are derived from the relationship between obligations and workload, both of which may vary from year to year in ways that are not necessarily correlated. As a result, service-level unit-cost and expenditure projections can vary substantially from base year to base year.

The methodology for projecting unit costs is different for different types of services. Table 3.2 shows the unit-cost basis for each service category included in the EHCPM. For durable medical equipment, prosthetics, and care provided through VA special programs, DSS provides unit-cost data that can be input directly into the UCPM and trended forward in time. For most other services, DSS costs are derived at too high a level to be input directly into the model (e.g., inpatient and ambulatory care). An aggregate relationship between VA and Medicare-allowable or billed charges is used to estimate VA unit costs at finer levels of service detail. Detailed unit costs are related either to Medicare-allowable charges in the base year or to community-billed charges (when the service is not covered by Medicare). The services for which unit costs are derived in this way constitute roughly 90 percent of the VA budget for modeled health care services. For a few services, the VA routinely purchases care from the private sector at community-billed charge levels (e.g., maternity care). For these services, VA unit costs are considered to be equal to community-billed charge levels.

The methodology used to derive detailed VA unit costs from detailed Medicare and community charges broadly corresponds to the method used to derive VA-specific utilization rates from HCG-based rates for the UPM. The derivation involves a series of adjustments to charge-based costs to assure that resulting unit costs reflect

Table 3.2
Unit-Cost Basis, by Health Care Service Category

Type of Service	Unit-Cost Basis	Detailed Health Service Category
Inpatient	Medicare-allowable charges	Medical
		Surgical
		Psychiatric
		Substance abuse
		Skilled nursing facility/extended care facility (non-acute)
	Community-billed charges	Maternity deliveries
		Maternity non-delivery
Ambulatory	Medicare-allowable charges	Allergy immunotherapy
		Allergy testing
		Anesthesia
		Cardiovascular
		Chiropractic
		Consults
		Emergency room
		Hearing and speech exams
		Immunizations
		Miscellaneous medical
		Office/home visits
		Outpatient psychiatric
		Outpatient substance abuse
		Pathology
		Physical exams
		Physical medicine
		Radiology
		Surgery
		Therapeutic injections
		Urgent care
		Vision exams
	Community-billed charges	Maternity deliveries
		Maternity nondelivery

Table 3.2—Continued

Type of Service	Unit-Cost Basis	Detailed Health Service Category
Pharmacy	Community-billed charges	Prescription drugs
	DSS direct	Over-the-counter medications
		Supplies
Miscellaneous	DSS direct	Glasses/contacts
		Hearing aids
	Community-billed charges	Ambulance
	DSS direct	Durable medical equipment
		Prosthetics
		VA program equipment and services
		Compensation and pension exams
Outpatient mental health programs	DSS direct	Day treatment
		Homeless
		Methadone
		Mental health intensive case management
		Work therapy
		Community residential care
Special VA programs	DSS direct	Blind rehabilitation
		Spinal cord injury
		Sustained treatment and rehabilitation
		Psychiatric residential rehab treatment
		Post-traumatic stress disorder residential rehab
		Substance abuse residential rehab treatment
		Compensated work therapy for the homeless chronic mentally ill
		Residential rehabilitation treatment

VA benefits, veteran demographics, projected management changes, and service-specific VA inflation and intensity trends. Relating VA unit costs to Medicare-allowable or community-billed charges enables the EHCPM to adjust VA unit costs for projected changes in health care management based on the unit-cost intensity relationships provided by the charge data.

VA unit costs are related to Medicare or community-billed charges in the base model year; however, the Medicare or community-billed charges are projected using VA trend assumptions. The model does not assume that projected VA unit-cost levels will trend in the same manner as Medicare or community-billed charges. For services for which unit costs are derived through this methodology, projected expenditures can be modeled in terms of Medicare or community-billed charges. The ability to compare projections using VA-based expenditures to those charge-based unit costs is useful for strategic-planning and policy analysis purposes. The following section describes this methodology in detail.

Derivation of Detailed Charge-Based Unit Costs. Here we provide a detailed description of the methodology used by Milliman to derive HSC-specific unit costs for services for which DSS cost information cannot be directly input into the UCPM. Table 3.3 outlines key steps in this process, extending the office-visit utilization example shown in Table 3.1. The process begins with the average Medicare-allowable charge for the intensity mix of office visits provided by the VA nationally in the base year (line *A*). This charge is calculated by multiplying Medicare-allowable charges per unit of service times the average number of relative value units (RVUs) associated with the service in the base year as documented in VA workload data.

Table 3.3
Derivation of Detailed Unit Costs Using Office-Visit Example

Label	Factor/Adjustment	Community Loosely Managed	Community Well Managed[a]
A	Medicare[b] national average (base year 2005)	$48.98	$48.98
B	HCG area adjustment	1.129	1.129
C	HCG WM area intensity adjustment	n/a	0.899
D	HCG covered-benefit adjustment	1.000	1.000
$E=A \cdot B \cdot C \cdot D$	Intermediate unit cost	$55.30	$49.70
			VA Specific
F	VA DoCM		22.00%
G (uses *E* and *F*)	Intermediate unit cost		$54.07
H	HCG age/gender adjustment		1.018
I	VA trend assumption—intensity		1.004
J	VA trend assumption—inflation		1.040
$K=G \cdot H \cdot I \cdot J$	Medicare[b] unit-cost benchmark		$57.47
L	VA/Medicare[b] cost relativity (base year 2005)		1.428
$M=K \cdot L$	VA projected unit cost		$82.07
$N=M \cdot o$[c]	VA total expenditures		$2,051.75

SOURCE: Adapted by RAND and ActMod from training materials provided by Milliman, Inc., and the VA.

NOTES: Cell specifications are benefit—office visit; fiscal year—2008; geographic sector—03-c-9-F; enrollee type—priority level 2, pre (non OIF/OEF), males, age category 55–59, age cohort 50–54 in 2005.

[a] While WM utilization rates are national, WM costs are community specific because of geographic differences in cost of care.

[b] Community-billed charges are used for services not covered by Medicare or when the VA purchases care at community-billed charge levels.

[c] Line *o* from Table 3.1.

Medicare-allowable charges are shown in line *A* and are the same for LM and WM plans because reimbursement for services with the same RVU value are established by CMS and do not vary within each geographic region. However, a number of adjustments are made to these national rates. First, they are adjusted to reflect differences in average Medicare-based unit costs due to regional differences in cost of living and practice patterns (line *B*). The unit costs for LM and WM plans in line *A* are the same because the relationship between the LM and WM charges is tied to the geographic area being modeled. Charges per service are distinguished by LM and WM as part of the geographic area adjustments. Geographic variations in the cost of delivering care related to geographic differences in practice costs (and not practice style) are accounted for in line *B*. Geographic differences between LM and WM plans in the intensity of care (measured by the number of RVUs per unit of service) due to management are accounted for in line *C*. An adjustment greater than 1.00 produces a WM unit cost that is greater than the LM unit cost and implies that WM results in a more intense service mix for a given unit of service. An adjustment less than 1.00 implies that WM results in a less intense service mix.

Next, unit costs are adjusted to reflect covered benefits (line *D*). However, this adjustment is applied for only a small set of services. These three adjustments result in local Medicare-allowable unit costs for services delivered under LM and WM plans (line *E*). The VA DoCM level (line *F*) is then applied to the local LM-WM interval in the same way that it is applied in the UPM described above to establish a base-unit cost for the local area (line *G*). Finally, unit costs are adjusted to reflect the intensity and mix of services specific to the modeled age/gender cell (line *H*) under the assumption that the relationship between unit costs and age and gender in the HCGs is comparable to that in VA settings.

The next step utilizes the relationship calculated between VA and Medicare (or billed) unit costs in the base year to develop a projected VA unit cost. Trended charge-based unit costs in each projection year (line *K*) are adjusted to reflect the base year relationship (line *L*) between VA unit costs (developed at a very aggregated service level) and charge-based unit costs to arrive at a VA-projected unit cost (line *M*). The derivation of trend factors is discussed below. In a final step, detailed unit costs are multiplied by projected utilization in line *o* of Table 3.1 to yield total projected office-visit expenditures.

Utilization and Cost Trends

For projections using the 2004 base year, trends were set based on a consensus developed among a panel of experts comprised of Milliman actuaries, senior VA leadership in consultation with program and field staff, and representatives from the White House Office of Management and Budget. While expert judgment still plays a central role, in more-recent versions of the model trend assumptions have been based on those anticipated by CMS and historical VA experience. As such, discussions with VA and Milliman staff indicated that panelists take pains to account (without double counting) for both various factors influencing changes in VA expenditures over time and VA data deficiencies, such as the lack of treatment capacity data.

Utilization Trends. The baseline morbidity- and reliance-adjusted VA utilization rate is adjusted to reflect assumptions about anticipated trends in utilization rates in the modeled projection year (Table 3.1, line *k*). Trend adjustments are applied to major groupings of modeled HSCs and can be either constant over multiple project years or specific to particular project years. Projections using the 2005 base year are fixed over multiple project years and assume that growth in the use of inpatient services will be flat or negative and utilization trends for outpatient care and prescription drugs are positive, reflecting expected changes in a range of influences, including practice patterns, technology, benefit design, population health status, patient preferences, and supply of services that influence the U.S. health care system generally. Changes over time in utilization due to reliance and community management practices are captured by their respective factors. For example, the model structure accounts separately for broad trends (e.g., inflation, intensity, utilization) by moving the entire WM-LM interval and trends in VA management that influence change in DoCM over time within the moving interval. Milliman also tests the reasonableness of individual trend components by measuring the implied aggregate trends in utilization and per-person, per-month expenditures.

Cost Trends. Milliman applies two types of trend adjustments to Medicare-based unit costs. The intensity trend (Table 3.3, line *I*) is intended to capture changes in the mix of services provided within an HSC over time as it influences unit cost. For some services, the intensity trend is adjusted for the ability of VA management to exercise some degree of control over the intensity of the service mix, which allows the VA unit costs to grow at a different pace than those in the private sector. Prescription drugs are an example of such a service. The inflation trend factor (Table 3.2, line *J*) captures changes in health care costs unrelated to intensity of care that influence the cost of care delivered in the VA and elsewhere. Inflation trend factors are intended to account for composite changes in the cost of employee compensation, drugs, supplies, equipment, and energy. As with utilization trends, cost trends are based on the judgment of VA officials in consultation with internal constituents and Milliman actuaries and are informed by the Medical Consumer Price Index and historic VA data.

The general structure of the trend components assumes that VA expenditures are similar to those generated under a fee-for-service system in which total expenditures are the product of utilization and externally established prices. Nonetheless, selection of year-specific trend factors can be used to incorporate changes in expenditure induced by changes over time in the VA's physical capacity relative to enrollee demand. For example, adjustments to utilization and unit-cost trends can be used to model the effect of supply constraints on expenditures. Likewise, increasing utilization and unit-cost trends can be used to model the effect of capacity expansions. Whether it is practical to adjust trend factors in this way depends on the availability to model developers of information about treatment capacity and the nature of unmet demand.

Budget Reconciliation

The VA's Allocation Resource Center provides Milliman with unit-cost data based on preliminary estimates of the total VHA obligation in the base year. In a final step, unit costs based on preliminary obligations are reconciled to the actual final obligation for the same base year. Differences in the preliminary and final obligations for FY 2004 can be traced to the modification of several budget items, such as collected fees, to the FY 2004 obligation.

CHAPTER FOUR

Findings on Model Structure and Validity

The team applied the criteria discussed in Chapter Two to evaluate the validity of the EHCPM, given the VA's policy context and objectives in sponsoring development of the model. The overall structure of the EHCPM is highly flexible and thus can accommodate a wide variety of potential specifications defined in terms of data inputs, service-level detail, geographic specificity, and enrollee type without major modifications. Because short-term forecasts are grounded in recent VA experience, we find that the model provides a valid and flexible platform for the purpose of near-term budget planning under the assumption that the current policy environment remains stable. At the same time, however, we find that the current specification of the ECHPM does not appear to yield demand-based forecasts of the effects of changes in the policy environment on VA resource requirements. Our concerns about the validity of current specification for policy planning stem from several sources:

1. The current specification does not adequately account for potentially important drivers of enrollment demand, such as changes over time in the generosity of private health insurance options available to veterans.
2. The current specification does not adequately account for potential confounding among enrollee case-mix, VA treatment capacity and reliance, and enrollees' preferences for treatment in VA facilities.
3. Utilization projections do not appear to be independent of the VA's capacity to deliver care.

4. The validity of the current specification relies on untested assumptions about the comparability of VA and commercial case-mix and management practices and about the variability of the VA's cost structure.

We did not undertake a formal review of VA data systems as part of our evaluation. In many instances, however, we trace the source of our concerns about model validity to gaps in data reported to us. These gaps have, thus far, prevented model developers from testing key assumptions and from structuring model components to reflect the processes through which VA expends resources in delivering care to its enrollees. Although the current specification can accommodate a wide range of modifications aimed at strengthening the ability of the model to support policy planning, in some cases, modifying assumptions about the nature of enrollee demand and the VA's cost structure may require substantial effort and the development of new analytic tools.

Enrollment Projection Model

The EPM uses 20-year projections of the future veteran population and current and historical data on the characteristics of enrolled veterans to project the characteristics of the future enrollee population in any given year by priority level, enrollee type, geographic area, and age/gender category. Our evaluation found that the methodology used to project future enrollment is reasonable and likely to yield accurate projections in a stable policy environment. At the same time, the current specification of the EPM appears to lack the specificity to inform explicit scenarios regarding the relationships among VA benefit generosity, other sources of health coverage, veterans' enrollment decisions, and enrollee health status.

Instead, the model uses historical relationships between selected sociodemographic characteristics and enrollment to project future enrollment in each priority level. Although the health insurance options available to individuals are determined to a considerable extent by their socio-demographic status (Doty and Holmgren, 2006; Doty and Holmgren, 2004; Zuvekas and Taliaferro, 2003; Hargraves, 2002; Shi, 2000), the model accounts for the effect of the other health insurance options available to veterans on VA enrollment in a very indirect way. Under the current structure, for example, it would be possible to model changes in enrollment rates for broadly defined groups (e.g., age category or priority level) expected to result from changes in the relative generosity of VA benefits. However, the current structure is not specific enough to allow the development of projections based on any explicit mapping of a relationship between specific features of benefit generosity (e.g., copays measured in dollar terms) and enrollment rates for specifically defined groups (e.g., veterans with employer-sponsored retiree coverage) that are consistent with the literature on health plan enrollment.

Such specificity is important to the extent that veterans select the health option perceived to offer the most advantageous combination of cost and quality given their economic circumstances (Strombom, Buchmueller, and Feldstein, 2002; Buchmueller and Feldstein, 1996; Abraham, Vogt, and Gaynor, 2006–2007). The studies referenced imply that it is through relative generosity that future changes in both the economic circumstances of veterans and the VA benefit will ultimately influence veterans' demand for enrollment. An important implication of these studies for the VA is that the VA benefit design can remain stable, but changes in benefit generosity and the premium cost of veterans' non-VA health benefit options can have a potentially large effect on enrollment demand. The new Medicare Part D prescription drug benefit is an example of such a change. Consistent with the literature, the availability of drug coverage through Part D may have the effect of making the VA a relatively less attractive option to individuals who previously had Medicare supplemental insurance that lacked drug coverage. In addition to changes in VA benefits and treatment capacity, other examples of changes in the VA policy environment that may be important include trends in the availability and generosity of employer-provided retiree health benefits and the effects of state-level initiatives, such as those enacted in Hawaii and Massachusetts, aimed at increasing access to health insurance (Blumberg et al., 2006; Lewin and Sybinsky, 1993).

Utilization Projection Model

The UPM derives cell-specific utilization rates by applying a complex series of adjustments to commercially based utilization benchmarks to make them specific to VA case-mix and local VA management practices. VA-specific utilization benchmarks are then calibrated to actual VA workload to anchor near-term projections to recent VA experience. We identified four features of the UPM that may limit the validity of the model's utilization projections: (1) the absence of data required to derive demand-based benchmarks independent of VA capacity, (2) the model's dependence on potentially unrealistic assumptions to derive morbidity and reliance adjustments for non–Medicare eligible enrollees, (3) a potential confounding between morbidity and reliance for all enrollees, and (4) the complexity of the adjustments required to "tailor" commercial utilization benchmarks to VA management practices and case-mix a complexity that does not give the model the specificity needed to predict the effect of changes in policy and practice environment in which the VA operates on resource requirements.

Derivation of VA-Specific Utilization Benchmarks

One of the key challenges in developing demand-based utilization projections is the limited utility of VA workload data. VA workload data are limited because they reflect only the VA portion of the total care demanded by enrollees, as well as the capacity of the VA to produce care. Although somewhat apprehensive about the proprietary nature of the Milliman HCGs used by the EHCPM, we initially considered the general concept of using utilization benchmarks from commercial experience to be a potentially innovative strategy for deriving VA-specific utilization benchmarks unconstrained by VA capacity. However, closer inspection of the method reveals that model projections remain constrained by VA capacity through the A/E adjustment. The reason for calibrating adjusted HCG benchmarks to actual VA experience through the A/E adjustment is to compensate for residual variation in utilization in the base year unexplained by model-based adjustments to benchmarked utilization. As a result, near-term projections are driven by VA experience. In the absence of a direct adjustment for variation in treatment capacity across market areas, this calibration imbeds utilization constraints into utilization projections in regions in which the VA's capacity to deliver care is constrained and results in some of the very characteristics that the use of the HCGs in the EHCPM was intended to avoid. We note, however, that the model structure itself is flexible enough to accommodate unconstrained demand-based utilization projections when and if data permit.

Morbidity and Reliance Adjustments for Non–Medicare Eligible Enrollees

A key challenge in adjusting commercial benchmarks to make them specific to the VA population is the confounding between morbidity and reliance. Reliance on VA care instead of other sources of care is largely driven by two factors. The first is enrollee preferences for VA care (notwithstanding the VA's ability to meet this demand). The second is the VA's capacity to meet enrollee demand. It is reasonable to expect morbidity to influence both factors. Milliman uses a complex series of ancillary analyses to derive separate morbidity and reliance adjustments. Linked Medicare claims and VA workload data support the development of appropriately controlled morbidity and reliance adjustments for Medicare-eligible VA enrollees. Morbidity and reliance adjustments for the under-65 population are complicated by the inability to fully observe morbidity for enrollees who are only partially reliant on the VA. In the version of the model reviewed by RAND, Milliman estimated under-65 morbidity based on the diagnoses of a subset of under-65 enrollees who reported being heavily reliant on the VA for their health care needs. This approach is valid only if morbidity and reliance are uncorrelated, an assumption that is unlikely to hold. Milliman has been exploring an alternate approach—imputing under-65 morbidity based on observed relationships between morbidity and reliance for veterans over the age of 65. This alternative approach rests on the assumption that relationships between morbidity and reliance for veterans under the age of 65 mirror relationships between morbidity and reliance for Medicare-eligible veterans. Again, this assumption is strong and not necessarily justified given the differences in the sources of health insurance coverage and in health status between the two populations (Newhouse, 1993). None of the approaches we reviewed or discussed with Milliman staff considered enrollee preference and VA capacity as distinctly different determinants of reliance. However, treating reliance as a composite factor relies on the implicit assumptions that enrollee demand will fully absorb capacity expansions and that capacity will expand to meet demand. The extent to which either assumption will be met in practice is not clear. These assumptions also make policy scenarios involving reliance less realistic. Milliman cited the lack of data describing VA capacity as a barrier to the development of a more realistic adjustment approach. Again, our concern relates more to data availability than the overall structure of the EHCPM, which can accommodate more-realistic reliance information.

Benchmarking to Community Management Practices

The development of VA-specific utilization benchmarks includes the DoCM—complex adjustments intended to account for the efficiency of local delivery systems measured relative to the efficiency of health plans operating in the geographic proximity of VA facilities. As discussed in Chapter Three, the derivation of the DoCM involves the placement of the local VA health care system along an interval anchored at 100 percent by a national utilization rate specific to health plans and hospitals that Milliman considers to be well managed and 0 percent for utilization rates that Milliman considers to be characteristic of loosely managed health plans operating in the same local area as the VA. In the case of outpatient care, the placement of the local VA along the LM-WM interval is based on the judgments of VA officials and Milliman actuaries about the extent to which local facilities have implemented Advanced Clinic Access, an initiative aimed at improving access to VA care. In the case of inpatient care, placement is informed by the results of analyses measuring the number of "avoidable days" occurring in VA facilities compared to those experienced by facilities in the geographic proximity of VA facilities.

The validity and utility of the DoCM adjustment are difficult to assess for a number of reasons. First, it is not possible to assess the relevance of the WM benchmarks to the VA delivery system. Milliman treats as proprietary both the identities of health plans and hospitals whose utilization rates are used in deriving the national WM benchmarks and the algorithm used in deriving the WM benchmarks. Thus, it is not possible to determine whether the criteria for defining the WM benchmarks described in model documentation were appropriately applied.

Second, the validity of the WM-LM benchmarks and the accuracy of comparisons relating VA facilities to the WM-LM benchmarks depend on the comparability of VA case-mix and national Medicare case-mix. Avoidable inpatient days for WM and LM plans and VA facilities are calculated by applying a formula that compares actual length of stay to predicted length of stay, controlling for case-mix for Medicare-eligible patients. Implicit in the formula are the assumptions that (1) the relationship between patient severity and length of stay is similar in the commercial and VA sectors and (2) efficiency benchmarks developed using Medicare data apply to care delivered to non–Medicare beneficiaries. These assumptions are more likely to be met if VA case-mix is similar to the national case-mix upon which the avoidable-days formula is based, the case-mix of under-65 patients is similar in the VA and the commercial sectors, and the relationship between case-mix and length of stay is similar for Medicare and non–Medicare beneficiaries. Evidence regarding the performance of "off-the-shelf" case-mix adjustment methods, of the type applied here, in VA patient populations is mixed (Rosen et al., 2001; Warner et al., 2004; Selim et al., 2006). However, the model documentation did not contain references to empirical assessments of comparability. While Milliman actuaries expressed confidence in the comparability of VA and commercial case-mix, we did not review results of empirical analyses demonstrating comparability.

Third, the validity of measuring the efficiency of inpatient mental health care and substance abuse treatment days using the metric of avoidable days is not well established in the clinical and payment policy literature. Work on the development of case-mix adjustment methods for psychiatric conditions and substance abuse disorders is ongoing (Sloan et al., 2006; Rosen et al., 2002; Ettner et al. 2001; Jencks, Goldman, and McGuire, 1985; Pincus and Goldman, 1985). In contrast to a wide variety of medical and surgical conditions treated in inpatient settings, no general consensus exists among policymakers, clinicians, and health services researchers regarding the ability to measure the efficiency of treatment for mental health and substance abuse conditions with administrative data (Sloan et al., 2006). The lack of evidence surrounding the validity of managerial efficiency measures for these services may reduce the credibility of the DoCM benchmarks with key constituents.

Finally, the DoCM adjustments are highly complex and thus were challenging for our project team to interpret and evaluate. The adjustments do not appear to capture essential features of the VA policy and practice environment. For example, DoCM adjustments peg the magnitude of efficiency gains over time to the local difference between the WM and LM utilization levels. We are not aware of policy initiatives or performance benchmarks aimed at evaluating the VA against community standards. In particular, materials describing the Advanced Clinic Access initiative (the implementation of which forms the basis of assumptions regarding changes in outpatient DoCM over time) reviewed by our team do not mention moving toward community norms as an explicit goal.

Unit Cost Projection Model

The UCPM produces a trended average unit cost for each service category by age/gender and market cell in each projection year. Unit costs are derived through the allocation of the VA budget obligation for modeled services in the base year to actual VA workload for those services in the base year. The validity of the unit cost projections produced by the model depends on the extent to which the VA's cost structure resembles that of a fee-for-service health insurer in practice. If the VA's cost structure is substantially different in practice, then misleading expenditure projections may result. Unfortunately, the sensitivity of expenditure projections to the model's characterization of the VA's cost structure is difficult to assess with available data. Alternative approaches that distinguish between fixed and variable costs may yield more accurate projections, but they would likely require significant investments in data and analytic expertise.

The current model structure does not relate average unit costs to utilization changes. This implies that incremental (or "marginal") costs are equal to average unit costs, which would only be true if the VA's nonconstruction costs are highly variable. This modeling approach is reasonable when the fixed costs of producing care are low (as is the case with pharmacy services by contrast to imaging and radiology services) or when the staff and physical capacity to produce additional units of care are always available (through contractors or leasing agreements, for example). If these conditions are not met, expenditure projections based on average unit costs will be biased. The extent of the bias will depend on the level of fixed costs associated with providing a particular service, will vary by system capacity, will vary over time, and will likely be larger in the short run when treatment capacity is relatively fixed. We expect, for example, that the current model does a reasonably accurate job of projecting pharmacy costs because variable medication costs account for the vast majority of expenditures. By contrast, we expect that projected expenditures on imaging and radiology services are relatively less accurate because of the high fixed costs associated with purchasing equipment required to provide these services.

The VA and Milliman indicated to the RAND project team that they are aware of the limitations of using average unit costs and that they use the current approach because VA lacks the tools necessary to estimate fixed and variable costs. The VA also indicated that it has taken steps to develop its ability to measure treatment capacity and intends to enhance the current EHCPM by developing a staffing model. The project team did not review such efforts as part of its evaluation.

A staffing model approach would estimate unit costs in a delivery system characterized by substantial fixed costs. Staffing models map resources required to staff, supply, and maintain a health care delivery system to meet anticipated demand. Unlike the average-unit-cost approach in the current model, a staffing model assumes that unit costs vary with utilization levels. Key features of a staffing model include

- identification of major expenditure components by type of service (e.g., salaries and benefits, purchased services, rents, administration, equipment, office supplies)
- unit cost measures for each type of expenditure derived by mapping utilization to expenditures by cost component
- measured relationships between resource use and utilization by cost component
- assumptions about what happens when capacity is reached (e.g., a timetable for implementing capacity expansions)
- trend measures based on the expenditure histories for major cost components.

Despite the potential advantages of a staffing model approach, Milliman and the VA suggested it was not a realistic option at the current time because the data needed to develop such a model are not readily available.

Prototype Staffing Model

In order to gain additional insight into the nature of the bias that might result if VA's costs are not in fact highly variable, our study team developed a small, prototype staffing model that projects office-visit expenditure for 1,000 enrollees based on the sample model cell discussed in Chapter Three (see Tables 3.1 and 3.3). We calibrated expenditures from the prototype staffing model in the base year to actual base year 2002 office-visit expenditures implied in this example.

The prototype models one unit of a single fixed-cost expenditure component. This single cost unit can be thought of as a single doctor, whose cost is fixed when he or she is a salaried, full-time employee. The structure of the staffing model requires us to specify explicitly what the VA's policy response will be when the capacity of the fixed-cost unit is reached.

We use our prototype to project expenditures under four different assumptions about the fixed-cost share of total costs and the ratio of enrollee demand to the available capacity of the fixed-cost unit. Ratio values less than 100 percent represent excess capacity, and values greater than 100 percent represent a capacity constraint that results in some demand being unmet. Table 4.1 shows assumption values for each of the four scenarios modeled using our prototype.

For each of the four scenarios, we consider two extreme policy responses to capacity constraints. The first response is to fund additional capacity when the capacity of the fixed-cost unit is reached. The second response is not to fund additional capacity at any time during the projection period. For each scenario, we compare staffing model projections to projections consistent with the current EHCPM for the same 1,000 enrollees that assume incremental costs are 100 percent variable.

Scenario 1

Figure 4.1 shows expenditure for the base case scenario when fixed costs comprise 25 percent of total costs and when enrollee demand is 95 percent of available capacity. The solid line traces projected office-visit expenditure levels for enrollees in the sample cell consistent with the current EHCPM, which assumes costs are 100 percent variable. The dashed lines trace two types of expenditure projections under two alternative policies for handling capacity constraints when they are reached. The base case scenario shows that when fixed costs are relatively low, the "100 percent variable" and the staffing model projections are reasonably similar until the capacity of the fixed-cost unit is reached in year five. After this time, capacity expansion policies drive expenditure projections. The staffing model projections start off lower

Table 4.1
Cost Structure and Capacity Assumptions Considered Using Prototype Staffing Model

	Demand/Available Capacity	
Fixed-Cost Share	**95%**	**105%**
25%	Scenario (base case)	Scenario 3
75%	Scenario 2	Scenario 4

Figure 4.1
Projected Office Visit Expenditures for Priority 2 Males Aged 50–59 Who Enrolled Pre–Eligibility Reform, by Model Type, for Four Fixed-Cost Share and Available-Capacity Scenarios

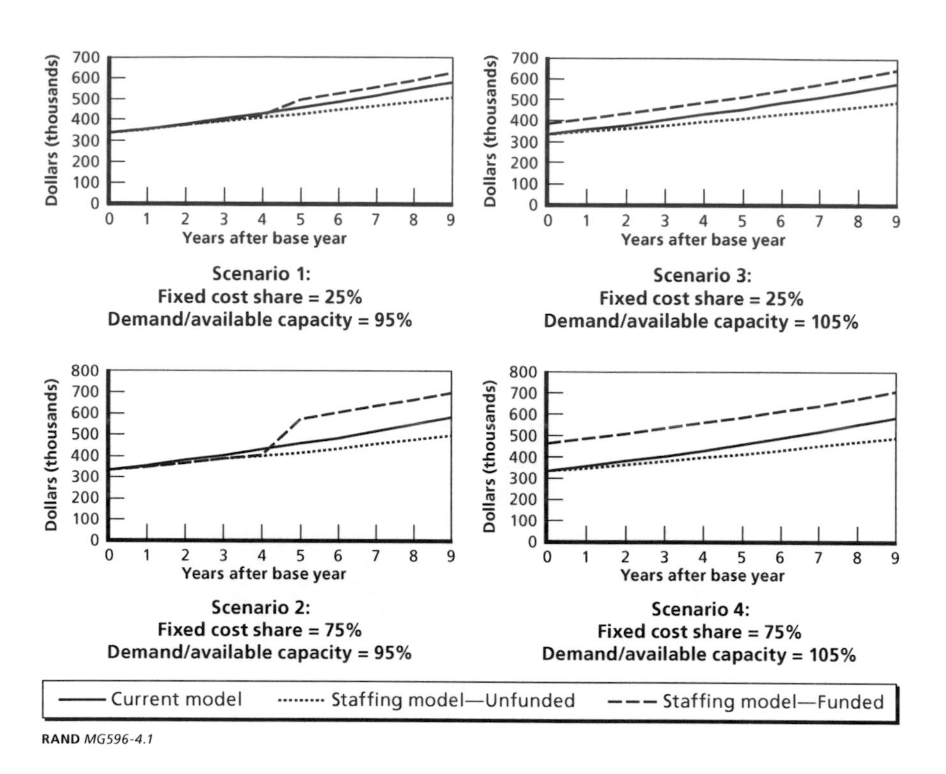

RAND *MG596-4.1*

because they reflect only the variable portion of incremental costs associated with increased utilization.

Scenario 2

Figure 4.1 shows expenditure projections when fixed costs comprise 75 percent of total costs and enrollees again demand 95 percent of available capacity in the base year. In a situation with excess capacity, higher fixed costs result in a larger difference compared to current model projections that assume incremental costs are 100 percent variable and to staffing model projections. When available capacity is reached, staffing model projections again depend on what we assumed about whether, when, and at what level capacity constraints will be funded.

Scenarios 3 and 4

In cases where capacity is strained in the base year, the current model and the staffing model are widely divergent in the base year, and the level of the divergence depends on the assumed cost of expanding the capacity of the fixed-cost unit required to meet excess demand.

To summarize, these scenarios illustrate the ways in which expenditure projections are sensitive to assumptions about the nature of fixed costs and policies regarding funding added capacity. It is not possible to know which approach is more accurate. Accuracy depends on which scenario more realistically captures relevant features of the VA cost structure. It is nonetheless worth highlighting the magnitude of the sensitivity, particularly when fixed costs are high. At the 3-year mark, the year most relevant for budget-planning applications, the difference under Scenario 2 between the staffing model and the current model was $10,907 for a single model cell comprised of 1,000 enrollees. Depending on base year utilization levels in different model cells, this difference could represent tens of millions of dollars in projected office-visit expenditures.

Utilization and Cost Trends

The current model specification includes five separate trend components: utilization, intensity, general inflation, reliance, and the efficiency of the VA's clinical management. To establish utilization, intensity, and inflation trend assumptions, Milliman conducts a trend analysis using an expert panel process informed by data describing recent trends in VA expenditures for selected cost components. Panelists include senior Milliman staff, senior VA leadership, and representatives from the White House Office of Management and Budget.

Our evaluation revealed no concerns about the expertise of the panelists and the process they use to formulate trend assumptions given the data that are currently available to them. At the same time, however, our evaluation raised several concerns about the specificity and validity of the trend factors as they are used in the model. First, the complexity of the trend factors does not increase the specificity of model projections. The linear relationship between trend factors (not including the DoCM trend) and expenditure projections means the detailed nature of the trend factors does not result in differential effects on projected expenditures. In other words, a 1-percent change in utilization trend assumptions will have the same effect on expenditure projections as a 1-percent change in the inflation assumptions.

Second, the complexity of the trend factors makes them challenging to verify. The EHCPM trend assumptions for treatment intensity are generally thought to be influenced by such factors as future advancements in medical technology and changes in providers' practices in coding the services they provide. Formulating trends assumptions based on these factors alone is a challenging task in and of itself. For example, using empirical data to forecast change over time in the intensity of service use (typically measured in RVUs) as distinct from changes in case-mix severity requires an extremely robust database and sophisticated case-mix adjustment methods. Distinguishing the impact of national trends from VA-specific trends (i.e., DoCM and reliance) in the absence of empirical data adds to the complexity of the task.

Third, the validity of cost-trend assumptions used in the model depends on the ability of experts to accurately adjust trend components derived from national fee-for-service experience to reflect the VA's cost structure and delivery system characteristics. To the extent that the VA has characteristics of an integrated delivery system, costs will be driven by major expenditure components, such as staff salaries and benefits, supplies, facility operation and maintenance, and medical equipment. In this circumstance, true unit costs will be driven by expenditures on each specific medical-care component (e.g., a medical equipment device) divided by the utilization served by that component. As the capacity and fixed cost of the component increases, the relationship between trend components and total expenditures becomes increasingly nonlinear and model assumptions relying on the direct proportionality of trend components and expenditures become less reasonable. Without data on treatment capacity, it is not possible to assess whether and at what point in time these relationships become nonlinear. Theoretically, careful adjustments of trend components related to treatment intensity and management efficiency can be made to minimize the bias created by using fee-service trends to forecast expenditures generated by an integrated delivery system. In practice, however, this is difficult without the detailed data on the components of total expenditures that would be required to actually specify a staffing model.

CHAPTER FIVE

Findings on Model Accuracy

The overall accuracy of the EHCPM remains uncertain. However, the accuracy of selected model components can be readily assessed. The VA assesses the accuracy of enrollment and patient projections regularly by comparing them to actual enrollment and actual numbers of patients, and it then provides the results of these assessments to internal stakeholders, including the Secretary of Veterans Affairs, and external stakeholders, including the Office of Management and Budget, congressional committees and staff, the Congressional Budget Office, and veterans service organizations.

Two factors complicate the assessment of the EHCPM's accuracy (beyond the factors that complicate the assessment of the accuracy of complex policy models generally). The first complication is the uncertain validity of key model assumptions. The validity of these assumptions plays a decisive role in helping to ensure a model's accuracy by reducing both the overall magnitude and systematic nature of forecasting errors. As we discuss above, the EHCPM treats (1) the VA case-mix and care processes as comparable to the commercial sector and (2) the VA's cost structure as variable. To the extent that the two sectors differ or costs are not variable in practice, the accuracy of the model is likely to suffer. Because these aspects of the model's structure remain largely untested, their impact on model accuracy is not known.

The second complication stems from the lack of unit-cost measures that are independent of the VA's budget allocation. The EHCPM strives, in essence, to hit a target that is not readily observable. Although it is tempting to assess the accuracy of expenditure projections used by the VA in budget planning by comparing 3-year projections to actual obligations for modeled services, such a comparison would be misleading. The VA's budget obligation reflects the results of political and strategic decisions, and the goal of the EHCPM is not to predict the outcome of these decisions. Best practices for assessing model performance in the context of budget planning require the development of a projection scenario that reconciles, at least in the aggregate, with the approved budget for an organization. For example, assume the VA submitted a budget request adequate to operate 1,000 clinics, but the ultimately approved budget amount only accommodated the operation of 950 such clinics. The VA would have to either find a way to operate the desired 1,000 clinics within the approved budget constraints or scale back the clinics' activities consistent with the approved budget. In either case, a projection scenario that reconciles with the approved budget (the "reconciled projection") should be prepared. If a reconciled projection were prepared, model accuracy could then theoretically be assessed by analyzing variances between the parameters in the reconciled projection (e.g., facility costs, unit costs, utilization, enrollment) and the corresponding actual parameters as experience emerges.

Under current specification, however, the derivation of reconciled projections is not feasible. This derivation is not possible because the current specification lacks an independent, resource-based unit-cost measure and, by extension, a resource-based measure of expenditures. Instead, unit costs are derived by allocating the VA's budget obligation—the VA's approved budget. Thus, expenditure projections and the approved budget are linked definitionally. As a result, expenditure projections are accurate by definition.

Whether unit-cost measures (derived by allocating the VA's budget obligation) accurately reflect resource requirements is also an open question. The development of a strategy for assessing accuracy is difficult because of the lack of an obvious gold standard against which to compare unit costs derived under the current specification. As noted in Chapter Four, unit costs for a staff model organization reflect the expenditures for defined resources (e.g. physicians, facilities, medical equipment) and the corresponding productivity and utilization of such resources. This contrasts with the EHCPM, in which categories of resources are not defined and projected unit costs are independent of projected utilization. It is not clear whether there are comparable delivery systems with which to compare the VA's cost basis.

Accuracy of the Utilization Projection Model

In the absence of an independent measure of expenditures, Milliman has appropriately focused its attention on assessing the accuracy of the UPM. Because the VA data systems upon which the model relies and the current structure of the VA benefit are both relatively new, opportunities for assessing accuracy against historical utilization experience have been limited. We reviewed documentation and results from a validation study conducted by Milliman under the direction of the VA. We discuss the results in greater detail in Appendix A. The study, referred to as the "BY02 Model Validation Study," assessed two sources of projection error between base year 2002 and forecast year FY 2003—forecast error and model error. Milliman defines *forecast error* as the difference between the projected change in utilization and the actual change in utilization. Forecast error occurs when trends and other delivery-system dynamics are inaccurately forecast. Milliman defines *model error* as the difference between modeled utilization (after A/E adjustments have been applied) and actual utilization in the base year. Model error occurs because A/E adjustments are applied at an aggregate level and not to the detailed age/gender and geographic-area cells. The validation shows 1-year forecast errors on the order of 0 percent to 9 percent and very negligible model errors across enrollee types for major service categories.

Although Milliman does not consider the size of the A/E adjustment to be a measure of model accuracy, the size of the A/E adjustment reflects Milliman's ability to adjust commercial experience to reflect VA case-mix and practice style. Small adjustments suggest a close relationship between benchmarked utilization and actual workload. In actuarial modeling applications, it is common to use analysis of the size and distribution of A/E adjustments across cells to inform model development and refinement. However, the lack of data on VA treatment capacity complicates this type of analysis. In the absence of treatment capacity data, it is not possible to determine whether large A/E adjustments arise from differences in case-mix or constrained capacity.

Quality Assurance Procedures

Computation errors and errors in mathematical algorithms used to derive projections can be a source of model error that, if undetected, can result in inaccurate projections. Discussion with the VA and Milliman staff suggests that Milliman has a number of systems in place to avoid and detect such errors when they occur. However, we did not review the specific features of Milliman's quality assurance process as part of our evaluation.

CHAPTER SIX

Findings on Tractability and Transparency

Tractability and transparency facilitate understanding of model structure and supporting data. These two features permit outside constituents to draw informed and independent conclusions about model quality and enable model staff to monitor quality and implement model enhancements. The modular design of the EHCPM's high-level structure is both tractable and transparent. However, the complexity of model subcomponents and the algorithms used to derive adjustment factors limit both tractability and transparency. Tractability and transparency are further reduced by the uneven and incomplete nature of the model documentation and by the use of proprietary model elements. This chapter discusses each of these issues in further detail.

High-Level Structure

The EHCPM is a flexible, component-based model. The model's three primary subcomponents—the utilization, enrollment, and cost models—each produce results that can be analyzed alone or combined to develop final projections. Because the subcomponents are combined through simple algebraic relationships (see the overview of model structure in Chapter Three), it is possible to substantially modify key elements of model subcomponents without having to modify the model's overall structure. Additionally, the model has a flexible structure that allows the user to substitute alternative assumptions without rewriting model code. For example, it would be possible to substantially modify the methodology used to develop each individual model subcomponent and yet maintain the model's overall structure.

Complexity

Complexity beyond that required to meet a model's primary objectives unnecessarily decreases tractability and transparency. While the high-level structure of the EHCPM is straightforward, the structure of model subcomponents and the ancillary analyses required to support model subcomponents are highly complex. This complexity is both expected and generally justified given both the uncertainties in the policy environment in which the VA operates and the gaps in the data available (e.g., reliance and morbidity) to develop factor adjustments and specify parameters. Under current policies, for example, demand for VA services is difficult to predict. Projecting demand under such circumstances typically requires multiple data sources, specialized analytic approaches, and numerous implicit and explicit assumptions. Enhancing validity and utility beyond current levels may require significant further increases in complexity.

Proprietary Model Elements

The use of proprietary elements in the utilization projection model creates an additional barrier to tractability and transparency, making the quality of an already complex model more difficult to assess. The HCGs and the methodology used to develop them are not available for review by the general public, and there is no formalized process to allow interested stakeholders and the general public access to HCGs and relevant supporting documentation. We were permitted to examine the HCG "rate book" during a two-day site visit to Milliman's Seattle, Wash., office. However, substantially greater access would have been required to understand the development methodology and the overall applicability of commercial utilization and performance benchmarks to the

VA delivery system, and to compare utilization and elasticity benchmarks to those in the published literature.

Documentation

Documentation is the primary mechanism through which model developers communicate with users (present and future) and constituents regarding the model's technical features. Effective documentation should accurately describe model features, list data sources, report key assumptions, and provide computational algorithms and equations. The EHCPM documentation, roughly 800 pages in total, covers the full breath of model features. However, incomplete and uneven descriptions of key model features made it considerably more difficult to understand the model's key features and assess model validity and utility.[1] Numerous typographical errors, inconsistent formatting, and the lack of an index further limited the documentation's usefulness.

---

[1] Throughout our evaluation, our project team sought to clarify the meaning of certain portions of the model documentation in multiple instances. In order to assure our accurate understanding of model structure and features, we submitted an early draft of Chapter Three, which presents an overview of the model, to our project sponsor and Milliman for comment and correction.

CHAPTER SEVEN

Benefits and Risks of the EHCPM

The VA asked RAND to consider the overall benefits and risks of relying on the EHCPM, as well as several specific issues related to the outsourcing of model development, maintenance, and expenditure projection through the current contractual arrangement. In addition to project management costs, the VA pays Milliman roughly $3.5 million per year for the actuarial services provided by 7.5 full-time-equivalent personnel who carry out the development, maintenance, and expenditure-projection tasks. The VA's expenditures for this contract constitute less than 0.02 percent of its $30 billion health care budget. While a quantitative assessment of the cost effectiveness of the EHCPM was beyond the scope of our evaluation, our evaluation was sufficient for us to draw conclusions regarding a broad range of benefits and risks to the VA posed by the current model specification and contractual arrangement. In particular, the VA asked RAND and ActMod to consider the risk of three specific scenarios: (1) Milliman goes out of business, (2) Milliman does not recompete, and (3) Milliman loses key staff. In addition, we discuss what we consider to be the most relevant risks of outsourcing generally—the loss of institutional knowledge generated by day-to-day engagement in the modeling process.

Benefits of the Current EHCPM

The current specification represents a substantial improvement over the VA's traditional methods for budget planning and performance monitoring. Like traditional VA methods, however, the current EHCPM specification is grounded in current-system capacity and thus is limited in its ability to inform budget scenarios beyond the status quo policy environment. The main potential costs associated with the EHCPM stem from risks to the VA's credibility and strategic-planning process resulting from inappropriately applying the model to inform scenarios beyond the current policy environment. Fortunately, the EHCPM also has a flexible architecture than can be modified and improved to expand the model's uses beyond its current capabilities as requisite data become available. We discuss each of these points in greater detail below.

The EHCPM offers improved budget-planning capabilities. The current model can support a wide array of budget-planning tasks because it (1) builds total expenditures from detailed service categories and enrollee types and (2) disaggregates enrollment, utilization, and cost components. These capabilities represent a substantial improvement over traditional budgeting methodologies. The VA can use the current specification to identify factors that drive expenditures and to develop more-informed strategies for managing expenditures and allocating budget appropriations. Also of interest to the VA is whether the model can support a variety of performance and budget monitoring tasks. Because the model can incorporate a wide range of assumptions about utilization and unit costs for a wide variety of services and enrollee types (e.g., special conflict—OIF/OEF—veterans, pre– and post–enrollment reform enrollees, and enrollees in capacity constrained markets), the current model structure can be used to monitor budget execution and performance relative to preestablished benchmarks. With respect to these two functions, it is the accuracy and timeliness of VA data systems, not the model's structure, that limits the utility of the model.

The current model also offers a limited capability to project the impact of "out-of-cycle" events (e.g., the enrollment of special-conflict veterans, the impact of hurricane Katrina). Using the model in this way is appropriate to the extent that any increased utilization resulting from these events can be accommodated with existing capacity, the period of consideration is in the relatively near term, and veterans enrolling as a result of the event under consideration are similar to current enrollees in terms of case-mix. For example, it must be the case that new priority 1 enrollees age 50–59 after these events are similar in their propensity to use health care services to current priority 1 enrollees age 50–59, unless a specific adjustment is made to the modeling factors (as is the case for OIF/OEF). In modeling out-of-cycle events under the current specification, it is also crucial to assess the appropriateness of the variable cost assumptions implied by the unit costing methodology. These preceding caveats notwithstanding, we also note that the VA has access to Milliman actuaries who can often adjust current model constraints to accommodate a VA-specific request.

The EHCPM offers a flexible platform for model enhancements. Our review identified concerns about the utility of the model for the purpose of strategic planning. As we discuss in the subsequent chapter, major modifications are required to support some of the VA's objectives in sponsoring the model. Because the EHCPM is a flexible, component-based model and because the subcomponents are combined through simple algebraic relationships, it is possible to substantially modify key elements of model subcomponents in no particular order without having to modify the model's high-level structure. Thus, under the model's current structure, the VA can continue to use the model as a budget-planning tool relatively undisrupted while model enhancements are implemented concurrently.

Risks of the Current EHCPM

The central role that Milliman plays in developing and maintaining the EHCPM is unique. While Milliman may assist other federal agencies, such as the Social Security Administration, CMS, and the VA Office of the Actuary, such agencies also rely heavily on their own internal actuarial staff. Outsourcing model-related activities is not necessarily inappropriate and may in fact be optimal. The appropriateness of outsourcing depends on what resources are available to the VA and what resources can be secured externally and at what cost. The materials we reviewed through the course of our evaluation did not permit us to evaluate the overall desirability of outsourcing in this unique situation.

However, we were able to draw conclusions about the three specific risks VA asked us to consider. We discuss each of these below.

Model-related activities are unlikely to be disrupted by Milliman's going out of business. Milliman is a multinational company with over $575 million in annual revenues. Milliman has been providing actuarial consulting services for over 50 years and is among the largest consulting and actuarial firms in the United States. Given that Milliman is a profitable and well-established company with a broad and diverse client base, we believe that it is unlikely that Milliman will go out of business in the foreseeable future. In the unlikely event that Milliman does go out of business, the availability and continuous maintenance of Milliman's HCGs could be interrupted. Although RAND and ActMod did not review specific contractual language, discussions with VA staff suggest that the VA owns the model logic and computer code "up to" the HCGs. While the VA's access to the HCGs might be interrupted if Milliman closed, projections from prior years could be used as a stop-gap until an appropriate alternative could be established. Alternatives include benchmarks maintained by other actuarial firms and the VA's own workload data. Should Milliman go out of business, the consultants engaged under the contract would potentially be available either individually or collectively to continue to support the VA EHCPM initiatives. Whether core Milliman staff could be retained by the VA in the event that Milliman closed would depend on whether the VA could establish requisite contracts in a timely manner.

Model-related activities are unlikely to be disrupted by Milliman opting not to recompete for the EHCPM contract. In our view, the failure of Milliman to recompete for the EHCPM contract represents greater risk than Milliman going out of business. However, we speculate that this risk is low given the size of the contract and its ongoing nature. Similar to a situation where Milliman goes out of business, Milliman's failure to recompete for the EHCPM would disrupt the VA's access to the HCGs. In this case, the VA could, as a stopgap measure, substitute benchmarks maintained by other actuarial firms. However, in this event, Milliman staff familiar with the model would not be readily available to continue to support model activities.

The loss of key Milliman personnel is more likely than Milliman going out of business. However, this risk is not unique to the VA's contract with Milliman and would exist under any contractual arrangement or under in-house production. However, replacing key project staff under an outsourcing arrangement is likely to be easier than replacing in-house production staff because staff with modeling and/or actuarial experience command high salaries, which may be difficult to meet under the Civil Service pay scale. There are currently two Milliman employees critical to the operation of the current contract. The first is a project director who oversees the day-to-day operation of the contract. The second is the principal actuary, who provides substantive expertise in overseeing the project. A sudden and unexpected departure by one of these two staff members in the midst of the budget preparation process would be particularly disruptive, but not fatal. We believe that the contract could be successfully managed by either staff member until a suitable replacement for the departing staff member could be identified and trained. We have no concerns about Milliman's ability to recruit qualified staff in the event that an internal candidate cannot be identified.

The most relevant risk to consider in outsourcing the EHCPM is the loss of institutional knowledge. Although the primary purpose of the EHCPM is to produce expenditure projections, the modeling process generates institutional knowledge in areas of strategic importance to senior VA leadership. The staff who develop and maintain the EHCPM have knowledge of the key policy concerns facing the VA, develop detailed knowledge of VA data systems, and interact with a wide range of experts who have substantive and technical knowledge of the VA and other health care delivery and financing systems. Under outsourcing arrangements, contractor staff members interact with other VA staff members in highly structured ways. Thus, contractor staff may be less available to mentor junior VA staff or to share insights informally.

CHAPTER EIGHT

Conclusions

Our evaluation of the EHCPM focused on its ability to provide accurate and timely projections of future demands on VA resources consistent with VA's budget and strategic-planning objectives. In conducting our evaluation, we reviewed key model features; assessed the validity, accuracy, tractability, and transparency of the model; and assessed the benefits and costs associated with the current specification and several specific aspects of the current contractual arrangement.

EHCPM Use in Short-Term Budget Projection

Because the model is calibrated to VA workload and budget, we conclude that the EHCPM yields reasonable short-term projections if VA treatment capacity and the policy environment in which the VA operates remain stable. The EHCPM improves on traditional budget-forecasting methods because it supports the separate projection of enrollment, utilization, and unit cost components of total expenditures and because it supports budget projections by detailed health service category and enrollee type.

EHCPM Use in Strategic Planning and Policy Analysis

At the same time, we identified concerns regarding the validity and accuracy of the current approach for projecting future expenditures under budget and policy scenarios beyond the VA's current capacity to provide care. Our first concern centers on the complex series of adjustments that Milliman makes to its commercial utilization benchmarks to reflect VA experience. To be successful, the adjustment methodology must disentangle the confounding among case-mix, clinical efficiency, and factors driving enrollee reliance on VA care. Reliance, in turn, is related to the VA's capacity to deliver care and meet enrollee preferences across sources of care. Thus, a robust adjustment methodology requires a wide variety of linkable data, including patient-level data documenting VA workload; non-VA utilization by VA enrollees; enrollee health, disability, and insurance status; and quantitative measures of VA's treatment capacity and clinical performance.

Our ability to assess the adequacy of the adjustment process was limited by the proprietary nature of the methodology used by Milliman in formulating the HCGs and clinical efficiency benchmarks. At the same time, however, the lack of adequate data on VA capacity and enrollee characteristics led us to question whether Milliman's adjustments to its commercial benchmarks considered the full range of the factors likely to be important for long-term planning.

Our second concern centers on the methodology used to derive unit costs. We found that Milliman's method of allocating the VA's approved budget to workload data to produce unit-cost measures would be accurate if the VA's nonconstruction costs are highly variable, as is the case for fee-for-service insurers such as Medicare. Resulting biases are likely to be most serious for services with large fixed-cost components for both capacity constrained markets and markets with substantial excess capacity. The current costing methodology also complicates any assessment of model accuracy because model projections are not independent of the VA's approved budget.

Future Modifications to the EHCPM

As a result of these concerns, we conclude that the current specification of the EHCPM has limited usefulness for policy planning. To enhance its usefulness for this purpose, the VA needs to develop analytic tools for measuring demand for health care, treatment capacity, and the fixed and variable costs associated with delivering care.

Forecasting the effects of VA policy and external influences on demand requires routine collection of data on veterans' employment, health insurance, health status, and overall health care utilization. Modifications to the VA's veteran surveys would allow for the collection of this more detailed information and the survey results could be linked to other VA data sources describing individual veterans' access to VA services (given local VA capacity), enrollment, and VA utilization. We did not determine the practical feasibility of collecting VA capacity data, but these data would be necessary for understanding the relationship between demand and capacity.

If, in fact, VA costs have a large fixed-cost component, then substituting the current UCPM with a staffing model approach to measuring unit costs may yield more-valid and more-accurate expenditure projections that can be more readily related to the VA's actual expenditures. A staffing model approach would build unit costs from the "bottom-up" from VA utilization and expenditures on major cost components, such as salaries and equipment, in the base year. Because unit costs derived from a staffing model would be independent of the VA's approved budget, the accuracy of model projections could be more readily assessed.

The implementation of a staffing model based on VA data would be a very time-consuming and resource-intensive activity involving a considerable investment in data collection. However, creating the capacity to develop, implement, and maintain a staffing model would most likely produce returns beyond the ability to improve the quality of model-based expenditure projections. Because a staffing model maps major expenditure categories to workload, it has the potential to inform the development and refinement of productivity benchmarks for physicians, physician support staff, and medical equipment, as well as the accurate measurement of performance relative to these benchmarks. A staffing model could also help the VA evaluate the potential return from investments in cost-saving or quality-enhancing technology.

These potential modifications to the EHCPM to enhance its role in policy planning would represent a significant investment in new analytic tools. If this is not practical or feasible, the VA may want to investigate simplifications to the current model for its use in short-term budget planning, drawing more exclusively from VA data sources and minimizing the use of commercial utilization benchmarks. A simpler model would be more transparent and may perform as well. We also conclude that the model-development process could be improved by better and more-complete documentation, the involvement of a wider range of expertise in model development, and a periodic review of model features and performance by independent experts.

APPENDIX A

Results of Validation Studies

BY02 Validation Study

Milliman conducted this study in order to measure the accuracy of utilization projections employed by the base year 2002 version of the EHCPM. The study compared modeled utilization in the base year to actual utilization in the base year by age, priority, service category, enrollee type, and VISN. These comparisons allow Milliman to identify cells in which modeled utilization is notably different from actual utilization and to attempt to modify the model or the level of the A/E adjustment to reduce observed discrepancies. The validation study explores two sources of projection error—forecast error and model error. *Forecast error* is the difference between the projected change in utilization and the actual change in utilization. Forecast error occurs when trends and other delivery system dynamics are inaccurately forecast. Validation study documentation provides several specific examples of the sources of forecast error:

- changes in health care trends
- health care management practices that did not meet expectations
- operational changes not considered in projections (e.g., staffing changes, policy changes, facility closures and openings, and benefit changes)
- operational changes considered in projections that had unexpected effects

- the supply of services that did not increase sufficiently to meet increased demand.

Milliman defined *model error* as the difference between modeled utilization (after A/E adjustments have been applied) and actual utilization in the base year. Under this definition, model error occurs because A/E adjustments are applied at an aggregate level and not to the detailed age, gender, and geographic-area cells. If A/E adjustments are applied at too fine a level of detail, there is a risk that projections will reflect random variation in utilization. If the A/E adjustment is performed at too aggregate a level, there is a risk that projections will not reflect important sources of variation in base year utilization that will influence use in future years. Model error may not have a big influence on the accuracy of near-term projections. However, as we note above, model error can become a substantial component of projection error to the extent that the system outgrows the A/E adjustments. Milliman assessed the potential of model error to contribute to forecast error in longer-term projections by (1) measuring the direction and magnitude of A/E factors and (2) comparing the relative value (or "slopes") of modeled utilization across detailed age categories and geographic areas.

Table A.1 shows A/E factors, model errors, and 1-year projection errors for selected service categories and enrollee types. This table shows a subset of health service categories that are particularly costly for the VA. The base year 2002 model projected that each of these services would have expenditures in excess of $1 billion in 2005. In addition, Table A.1 shows results for priority 1 and priority 5 veterans, the two groups with the highest expenditures. As can be seen in columns 1 and 4 of the table, there are frequently large discrepancies between actual and modeled values in the base year before the A/E adjustment is applied. For example, Table A.1 shows that the A/E factor for priority level 1 "pre" enrollees under the age of 65 for pathology was 1.64, implying that actual utilization was 64 percent higher than utilization predicted by the adjusted HCGs after they are adjusted for geography, case-mix, and management efficiency but not yet by the A/E factor. The model error ratio, shown in columns 2 and 5, compares modeled

Table A.1
Actual-to-Expected Adjustment Factors, Model Error Ratios, and Projection Error Ratios by Selected Service Types and Selected Enrollee Types

	Under 65			65 and Older		
Type of Service and Enrollee Type	A/E Factor	Base Year 2002 Model Error Ratio	FY 2003 Projected Error Ratio	A/E Factor	Base Year 2002 Model Error Ratio	FY 2003 Projected Error Ratio
Priority level 1—Pre						
Inpatient medical	0.94	1.00	0.93	0.86	1.00	0.96
Inpatient surgical	1.22	1.00	1.01	1.24	1.00	0.99
Pathology	1.64	1.00	1.01	1.43	1.00	1.01
Radiology	1.08	1.00	1.08	0.68	1.00	1.00
Office visits	1.10	1.00	1.08	0.93	1.00	1.08
Ambulatory surgery	0.86	1.00	1.00	0.83	1.00	1.01
Prescription drugs	0.93	1.00	1.00	0.89	1.00	1.04
Priority level 1—Post						
Inpatient medical	0.79	1.00	0.97	1.01	1.00	1.00
Inpatient surgical	1.13	1.00	1.08	1.38	0.98	0.98
Pathology	1.82	1.00	0.95	1.47	1.00	0.98
Radiology	1.80	1.00	1.04	0.86	1.00	1.04
Office visits	1.12	1.00	1.01	0.88	1.00	1.04
Ambulatory surgery	1.20	1.00	0.92	0.85	1.00	0.94
Prescription drugs	0.72	1.00	0.91	0.60	1.00	0.98
Priority level 5—Pre						
Inpatient medical	0.84	1.00	1.08	0.96	1.00	1.04
Inpatient surgical	0.99	1.00	1.07	1.20	1.00	1.01
Pathology	1.25	1.00	1.01	1.51	1.00	1.04

Table A.1—Continued

	Under 65			65 and Older		
Type of Service and Enrollee Type	A/E Factor	Base Year 2002 Model Error Ratio	FY 2003 Projected Error Ratio	A/E Factor	Base Year 2002 Model Error Ratio	FY 2003 Projected Error Ratio
Radiology	0.82	1.00	1.00	0.69	1.00	1.04
Office visits	0.71	1.00	1.07	0.83	1.00	1.09
Ambulatory surgery	0.57	1.00	1.04	0.73	1.00	1.05
Prescription drugs	0.77	1.00	1.04	1.11	1.00	1.08
Priority level 5—Post						
Inpatient medical	0.85	1.00	1.06	0.93	1.00	1.09
Inpatient surgical	1.05	1.00	1.03	1.23	1.00	1.02
Pathology	1.36	1.00	0.98	1.44	1.00	1.02
Radiology	1.16	1.00	1.00	0.68	1.00	1.03
Office visits	0.76	1.00	1.05	0.75	1.00	1.08
Ambulatory surgery	0.76	1.00	1.01	0.64	1.00	1.00
Prescription drugs	0.72	1.00	0.99	0.93	1.00	1.06

SOURCE: CACI, Inc., and Milliman, Inc., "Model Validation Report," Contract #GS-0F-226K, Task #101-C37055, Tables 7-1-b, 7-5-b, 7-9-b, 7-13-b, 7-17-b, 7-21-b, 7-25-b–7-29-b.

NOTES: Service categories included in this table constitute all services with projected expenditures greater than $1 billion in FY 2005; this comprises roughly 68 percent of all projected expenditures.

[a] Priority-level 1 and 5 enrollees consume the largest share of VA health care expenditures.

utilization after the A/E adjustment to actual utilization in the base year. By definition, the model error ratios are almost always equal to 1. Model errors not equal to 1 reflect situations where a single A/E adjustment factor was applied to multiple service categories.

Comparisons of actual-to-expected VISN relativities for selected services are presented in Table A.2. A VISN's relativity is defined as the ratio of the VISN-specific utilization rate to the VA-wide utilization rate for a specific service category. The columns under the various service categories (i.e., inpatient medical, inpatient surgical, office visits, and prescription drugs) show the ratio of modeled VISN relativities (i.e., the HCG-defined area factors) to the actual VISN relativities observed in base year 2002. Values close to 1 suggest that HCG area factors closely track the relativities observed in actual VA workload data. Age-based relativities are calculated and interpreted in a similar fashion and for base year 2002 are shown in the left-hand columns under each service category heading in Table A.2. Comparable figures for base year 2004 are shown in the corresponding right-hand columns.

The study documentation we reviewed did not include a formal assessment of the size and distribution of modeling errors or relativities compared to preestablished tolerance criteria. Nonetheless, the documentation traces patterns of large discrepancies between actual and modeled values to several factors: (1) unmeasured capacity constraints; (2) coding differences between the VA and HCG data that influence the comparability of utilization counts across the two systems; (3) geographic differences in the "mission" of VA facilities that reduce the comparability of private-sector based benchmarks; (4) geographic differences in the provision of special VA services resulting in demand patterns that the model does not capture; (5) confounding of age, morbidity, and reliance not captured by the model; and (6) demand effects of changes in VA treatment patterns, such as shifting care from more-intense to less-intense modalities.

Based on these results, Milliman hypothesized that the lack of model fit for 65-and-over age categories in base year 2004 resulted from reductions in reliance that occur as VA enrollees integrate into Medicare. Based on this hypothesis, Milliman used linked VA workload and Medicare data to age-adjust reliance factors for VA enrollees age 65 and older in the base year 2005 model.

Table A.2
Base Year Modeled to Actual Utilization Relativities by VISN

VISN	Inpatient Medical	Inpatient Surgical	Office Visits	Prescription Drugs
1	0.95	0.94	0.94	1.11
2	1.06	0.99	0.86	1.02
3	0.93	0.91	1.04	1.39
4	1.16	1.30	1.01	0.93
5	0.98	1.14	0.86	1.02
6	1.04	0.98	0.96	0.91
7	0.90	0.90	1.05	1.02
8	1.04	1.09	0.98	0.92
9	1.17	1.26	1.00	1.06
10	1.03	1.15	0.78	1.02
11	0.90	1.01	0.98	0.93
12	0.89	1.04	0.91	0.99
15	0.92	0.86	1.10	0.86
16	1.01	1.09	1.10	1.01
17	0.96	0.96	1.08	0.95
18	1.08	1.06	1.04	0.95
19	1.06	0.91	1.14	0.98
20	1.03	0.85	1.10	0.99
21	0.85	0.70	0.99	1.08
22	0.92	0.88	0.98	1.09
23	1.07	0.93	1.07	0.97

SOURCE: CACI, Inc., and Milliman, Inc., "Model Validation Report," Contract #GS-0F-226K, Task #101-C37055, Tables 5-1, 5-2, 5-31, and 5-43.

NOTE: VISNs 13 and 14 were incorpoated into neighboring VISNs and no longer exist.

Pharmacy Copay Utilization Impact Study

Despite substantial changes in enrollment policy, the VA benefit has remained virtually unchanged during the lifespan of the EHCPM. On February 1, 2002, the VA pharmacy copay for a 30-day supply of prescription medication was increased from $2 to $7. This change represented an opportunity to assess the accuracy of the EHCPM in predicting the effect of the copay change on the use of prescription drugs.

The assessment of model accuracy followed a two-step process. Milliman first used actual VA workload data to estimate the effect of the increased copay on the use of prescription drugs separately for veterans over and under age 65 using a pre-post comparison methodology. Milliman measured the effect by comparing utilization occurring pre– and post–copay change for priority 7c and 8c enrollees to the pre-post change for priority 1 enrollees who are not subject to copays and are thus unaffected by the copay change. The effect of the copay changes on priority 2 though 7a/8a was not examined because veterans in those groups are not charged for certain types of services and conditions. By contrast, priority 7c and 8c veterans were subject to copays on all prescriptions.

Milliman estimated that, compared to their priority 1 counterparts, groups with higher copays reduced prescription drug use by 5.3 percent among the priority 7c/8c veterans under age 65 and 6.0 percent among veterans 65 and older. Next, Milliman compared these estimates to the reduction in utilization predicted by the HCG-based elasticity benchmarks. The relative reduction predicted by the HCG benchmarks for two groups facing comparable benefit structures was 4.7 percentage points for both age groups. The HCG-based estimate was less than 1.5 percentage points different from the estimated impact based on actual utilization.

The small magnitude of this difference suggests that the model (more specifically, the HCG-based copay adjustment factor) was accurate in predicting the relative magnitude of the impact of the copay change. The HCG estimate was also slightly more conservative, which is consistent with the idea that demand for prescription drugs by VA enrollees is more elastic than demand by commercial enrollees.

APPENDIX B

Priority-Level Definitions

The following definitions were taken from information materials publised by the VA (U.S. Department of Veterans Affairs, 2008a, and U.S. Department of Veterans Affairs, 2008b).

Group 1: Veterans with service-connected disabilities rated 50 percent or more and/or veterans determined by VA to be unemployable due to service-connected conditions.

Group 2: Veterans with service-connected disabilities rated 30 or 40 percent.

Group 3: Veterans with service-connected disabilities rated 10 and 20 percent, veterans who are former prisoners of war or were awarded a Purple Heart, veterans awarded special eligibility for disabilities incurred in treatment or participation in a VA vocational rehabilitation program, and veterans whose discharge was for a disability incurred or aggravated in the line of duty.

Group 4: Veterans receiving aid and attendance or housebound benefits and/or veterans determined by VA to be catastrophically disabled. Some veterans in this group may be responsible for copays.

Group 5: Veterans receiving VA pension benefits or eligible for Medicaid programs, and non–service connected veterans and noncompensable, 0-percent service-connected veterans whose annual income and net worth are below the established VA means-test thresholds.

Group 6: Veterans of the Mexican border period or World War I; veterans seeking care solely for certain conditions associated with exposure to radiation or exposure to herbicides while serving in Vietnam; veterans seeking care for any illness associated with combat service in a war after the Gulf War or during a period of hostility after November 11, 1998; veterans seeking care for any illness associated with participation in tests conducted by the Defense Department as part of Cold War–era chemical weapons tests and Project "Shipboard Hazard and Defense;" and veterans with 0-percent service-connected disabilities who are receiving disability compensation benefits.

Group 7: Veterans with income and/or net worth above the VA established threshold and income below the HUD geographic index who agree to pay copays. Subpriority a: Noncompensable 0-percent service-connected veterans. Subpriority c: Non–service connected veterans.

Group 8. Veterans with income and/or net worth above the VA-established threshold and the HUD geographic index who agree to pay copays. Subpriority a: Noncompensable 0 percent service-connected veterans enrolled as of January 16, 2003, and who have remained enrolled since that date. Subpriority c: Non–service connected veterans enrolled as of January 16, 2003, and who have remained enrolled since that date.

APPENDIX C

Model Uses Described in "VA Enrollee Health Care Projection Model Training Companion Manual," June 2006

- Enrollment-level decision analysis
  - Projecting service utilization and associated expenditures
- Tracking enrollment changes and trends over time
- CARES baseline and analyses
  - Data generated from these models was used to begin the CARES evaluation process.
  - CARES was the driving force behind the development of unmet demand projections.
- Enrollee cost-sharing analyses
  - The model tests sensitivity to various copay and deductible scenarios.
  - It analyzes the impact changes to the copay structure will have on utilization and expenditures.
  - It also analyzes the impact of initial-use fees and other cost-sharing mechanisms.
- Budget formulation
  - Expenditure projections used to develop budget needs for upcoming years.
- Market and unmet demand analyses
  - Local-level data can be used to analyze market shares.
  - Unmet-demand analyses can be used to understand the impact of improving access.
- Planning model for VISNs

  - Treating facility demand can be used to determine if supply of services will meet demand for services.
  - Capital planning
  - CBOC planning
- Scenario testing
  - Answer "What If" questions regarding enrollment and expenditures
- Policy-decision analyses
  - Estimate impact of policy decisions on enrollment, expenditures, and revenue
- VERA funding allocations (potentially)
  - Use 20-year projections to estimate future VERA funding
  - Use modeled expenditures to allocate VERA funding to facilities within a VISN
  - Use modeled expenditures to estimate transfer payments
- Special analyses
  - Millennium Bill analyses
  - DoD analyses
  - VA Special-Program analyses
  - Reasonable-charges analyses
- Private-sector contracting
  - Model can be produced with Medicare-allowable charges (many private-sector provider contracts are a percentage of Medicare allowable)
  - Reasonableness of capitation contracts with private-sector providers can be assessed
  - Actual VA workload can be incorporated for contract pricing

References

Abraham JM, Vogt WB, Gaynor MS. How do households choose their employer-based health insurance? *Inquiry*. 2006–2007 Winter;43(4):315–332.

Actuarial Standards Board. *Appraisals of Casualty, Health, and Life Insurance Businesses*. Actuarial Standards of Practice No. 19, Doc. No. 099, June 2005.

Angist JD. Lifetime earnings and the Vietnam era lottery draft: Evidence from Social Security administrative records. *Am Econ Rev*. 80(3):312–336.

Agha Z, Lofgren R, VanRuiswyk J, Layde P. Are patients at Veterans Affairs medical centers sicker? A comparative analysis of health status and medical resource use. *Arch Intern Med*. 2000 November;160:3252–3257.

Asch SM, McGlynn EA, Hogan MM, Hayward RA, Shekelle P, Rubenstein L, Keesey J, Adams J, Kerr EA. Comparison of quality of care for patients in the Veterans Health Administration and patients in a national sample. *Ann Intern Med*. 2004 December 21;141(12):938–945.

Blumberg LJ, Holahan J, Weil A, Clemans-Cope C, Buettgens M, Blavin F, Zuckerman, S. Toward universal coverage in Massachusetts. *Inquiry*. 2006;43(2):102–121.

Booth, R, Chadburn R, Cooper D, Haberman S, James D. *Modern Actuarial Theory and Practice*. Boca Raton, Fla.: Chapman and Hall/CRC Press; 1999.

Buchmueller TC, Feldstein PJ. Consumers' sensitivity to health plan premiums: Evidence from a natural experiment in California. *Health Aff* (Millwood). 1996 Spring;15(1):143–151.

CACI, Inc., and Milliman, Inc., "Model Validation Report," Contract #GS-0F-226K, Task #101-C37055.

Cohany, S. The Vietnam-era cohort: Employment and earnings. *Mon Labor Rev*. 1992 June;115(6):3–15.

Congressional Budget Office. *The Potential Costs of Meeting Demand for Veterans' Health Care*. Washingon, D.C.; March 2005.

Doty MM, Holmgren AL. Unequal Access: Insurance Instability Among Low-income Workers and Minorities. *Commonwealth Fund Issue Brief.* April 2004 (729):1–6.

———. Health Care Disconnect: Gaps in Coverage and Care for Minority Adults. Findings from the Commonwealth Fund Biennial Health Insurance Survey (2005). *Commonwealth Fund Issue Brief.* August 2006 (21): 1–12.

Ettner SL, Frank RG, McGuire TG, Hermann RC. Risk adjustment alternatives in paying for behavioral health care under Medicaid. *Health Serv Res.* 2001 August;36(4):793–811.

Etzioni DA, Yano EM, Rubenstein LV, Lee ML, Ko CY, Brook RH, Parkerton PH, Asch SM. Measuring the quality of colorectal cancer screening: The importance of follow-up. *Dis Colon Rectum.* 2006 July;49(7):1002–1010.

General Accounting Office. *Guidelines for Model Evaluation* (PAD-79-17). Washington, D.C.; 1979.

———. *VA's Management of Drugs on Its National Formulary* (GAO/HEHS-00-34). Washington, D.C.; 1999.

Government Accountability Office. *Medicare Trust Funds Actuarial Estimates: Efforts Have Been Made to Improve Internal Control Over Projection Process but Some Weaknesses Remain* (GAO-03-247R). Washington, D.C.; March 4, 2003.

Gass SI, Thompson BW. Guidelines for model evaluation: An abridged version of the US General Accounting Office exposure draft. *Oper Res.* 1980 March–April;28(2):431–439.

Glied S, Remler DK, Zivin JG. Inside the sausage factory: Improving estimates of the effects of health insurance expansion proposals. *Milbank Q.* 2002;80(4):603–635, iii.

Hargraves JL. The insurance gap and minority health care, 1997–2001. Center for Studying Health Systems Change Tracking Report, 2002;June(2):1–4.

Inglehart, JK. Reform of the Veterans Affairs health care system. *N Engl J Med.* 1996 October 31;335(18):1407-1411.

Jencks SF, Goldman HH, McGuire TG. Challenges in bringing exempt psychiatric services under a prospective payment system. *Hosp Community Psychiatry.* 1985 July;36(7):764-9.

Kronick R, Gilmer T, Dreyfus T, Lee L. Improving health-based payment for Medicaid beneficiaries: CDPS. *Health Care Finance Rev.* 2000 Spring;21(3):29–64.

Lewin JC, Sybinsky PA. Hawaii's employer mandate and its contribution to universal access. *JAMA.* 1993;269(19):2538–2543.

National Research Council. *The Use of Microsimulation Methodology,* Volume 1: *Review and Recommendations*, Washington, D.C.: National Academy Press; 1991.

Newhouse JP, Insurance Experiment Group. *Free for All? Lessons from the RAND Health Insurance Experiment.* Cambridge, Mass.: Harvard University Press; 1993.

Old-Age, Survivors, and Disability Insurance Trustees. *Annual Report of the Board of Trustees of the Federal Old-Age and Survivors Insurance and Disability Insurance Trust Funds.* May 1, 2006.

Pincus HA, West J, Goldman H. Diagnosis-related groups and clinical research in psychiatry. *Arch Gen Psychiatry.* 1985 June;42(6):627–629.

Quade, ES. *Analysis for Public Decisions.* New York: North-Holland; 1989.

Remler DK, Zivin JG, Glied SA. Modeling health insurance expansions: Effects of alternate approaches. *J Policy Anal Manage.* 2004 Spring;23(2):291–313.

Rosen AK, Loveland SA, Anderson JJ, Hankin CS, Breckenridge JN, Berlowitz DR. Diagnostic cost groups (DCGs) and concurrent utilization among patients with substance abuse disorders. *Health Serv Res.* 2002 August;37(4):1079–1103.

Rosen AK, Loveland S, Anderson JJ, Rothendler JA, Hankin CS, Rakovski CC, Moskowitz MA, Berlowitz DR. Evaluating diagnosis-based case-mix measures: How well do they apply to the VA population? *Med Care.* 2001 July;39(7):692–704.

Savoca E, Rosenheck R. The civilian labor market experiences of Vietnam-era veterans: The influence of psychiatric disorders. *J Ment Health Policy Econ.* 2000;3(4):199–207.

Selim A, Kazis L, Rogers W, Qian S, Rothendler J, Lee A, Ren X, Haffer S, Mardon R, Miller D, Spiro A, Selim B, Fincke B. Risk-adjusted mortality as an indicator of outcomes: Comparison of the Medicare advantage program with the Veterans' Health Administration. *Med Care.* 2006 April;44:4.

Shi, L. Vulnerable populations and health insurance. *Med Care Res Rev.* 2000;57(1):110–134.

Sloan KL, Montez-Rath ME, Spiro A, Christiansen CL, Loveland S, Shokeen P, Herz L, Eisen S, Breckenridge JN, Rosen AK. Development and validation of a psychiatric case-mix system. *Med Care.* 2006 June;44(6):568–580.

Stineman MG, Ross RN, Hamilton BB, Maislin G, Bates B, Granger CV, Asch DA. Inpatient rehabilitation after stroke: a comparison of lengths of stay and outcomes in the Veterans Affairs and non-Veterans Affairs health care system. *Med Care.* 2001 February;39(2):123–137.

Strombom BA, Buchmueller TC, Feldstein PJ. Switching costs, price sensitivity and health plan choice. *J Health Econ.* 2002 January;21(1):89–116.

Stokey E, Zeckhauser R. *A Primer for Policy Analysis.* New York: Norton and Company; 1978.

U.S. Department of Veterans Affairs. "VA Health Care: Enrollment Priority Groups," Fact Sheet 164-2, March 2008a. As of September 15, 2008:
http://www.va.gov/healtheligibility/Library/pubs/EPG/

———. "Federal Benefits for Veterans and Dependants," Web page, March 11, 2008b. As of September 15, 2004:
http://www1.va.gov/opa/IS1/1.asp

Warner G, Hoenig H, Montez M, Wang F, Rosen A. Evaluating diagnosis-based risk-adjustment methods in a population with spinal cord dysfunction. *Arch Phys Med Rehabil.* 2004 February;85(2):218–226.

Zuvekas SH, Taliaferro GS. Pathways to access: Health insurance, the health care delivery system, and racial/ethnic disparities, 1996–1999. *Health Aff,* 2003;22(2):139–153.